SCIENCE AND HUMANITY

SCIENCE AND HUMANITY

A Humane Philosophy of Science and Religion

Andrew Steane

OXFORD
UNIVERSITY PRESS

Great Clarendon Street, Oxford, OX2 6DP,
United Kingdom

Oxford University Press is a department of the University of Oxford.
It furthers the University's objective of excellence in research, scholarship,
and education by publishing worldwide. Oxford is a registered trade mark of
Oxford University Press in the UK and in certain other countries

© Andrew Steane 2018

The moral rights of the author have been asserted

First Edition published in 2018

Impression: 1

All rights reserved. No part of this publication may be reproduced, stored in
a retrieval system, or transmitted, in any form or by any means, without the
prior permission in writing of Oxford University Press, or as expressly permitted
by law, by licence or under terms agreed with the appropriate reprographics
rights organization. Enquiries concerning reproduction outside the scope of the
above should be sent to the Rights Department, Oxford University Press, at the
address above

You must not circulate this work in any other form
and you must impose this same condition on any acquirer

Published in the United States of America by Oxford University Press
198 Madison Avenue, New York, NY 10016, United States of America

British Library Cataloguing in Publication Data

Data available

Library of Congress Control Number: 2018931548

ISBN 978–0–19–882458–9
DOI: 10.1093/oso/9780198824589.001.0001

Printed and bound by
CPI Group (UK) Ltd, Croydon, CR0 4YY

Links to third party websites are provided by Oxford in good faith and
for information only. Oxford disclaims any responsibility for the materials
contained in any third party website referenced in this work.

To Neil and Alison

Acknowledgements

I find myself filled with thankfulness to many people who have helped with this book.

First I would like to acknowledge all those who have written cogently and thoughtfully in the areas I have tried to present; many are authors of works included in the bibliography.

More directly, I would like to thank Sonke Adlung at Oxford University Press for his help in getting this book underway and on to publication. I also thank Heather McCallum who gave a helpful editorial steer on an earlier project; this book is part of the result. I thank Prof. P. Ewart and members of the Oxford Forum for insightful discussion and encouragement. I thank Dr J. Grant and Prof. R. Swinburne for their expert advice, and also the anonymous referees at OUP who offered many detailed points of constructive criticism. They were generous with their time and expertise and I am truly grateful.

I thank Prof. Derek Stacey, Prof. Harvey Brown, Dr Bethany Sollereder, Dr Noah Carl, and Dr Sally Hodgson who each read an early draft of one or more chapters and gave well-informed feedback. I thank Will Kemp for his cogent feedback on a set of poems which included most of those included here. I thank also my family for their constant and kind support, and Prof. Hans Halvorson for encouragement and some expert advice.

Finally, my special thanks go to Marco Bodnar, Caspar Jacobs, and Dr Phil Marriott for their generous help and thoughtful comments. Marco showed me some places where I was obscure; Caspar noticed where relevant considerations had been missed; and Phil, in a few wise words, deftly showed me the need to reconfigure the flow of the book. All this help should not be taken to imply complete agreement; this was a case of productive discussion and these friends each have their own independent viewpoints.

All these interactions helped me to see where my arguments were unclear, or insufficiently justified, or poorly expressed. I am very grateful for having such defects detected, and in consequence the book was much improved. The defects that remain are my own responsibility.

Counterpoint, Frequencies copyright © Kunjana Thomas 1988, 1990, 1992, 1995, 2004; extracts reproduced with permission of the Licensor through PLSclear.

The Cartographer Tries to Map a Way to Zion copyright © Kai Miller 2014, extract reproduced with permission of the Licensor and Carcanet Press Limited.

Lines from *R. S. Thomas, Selected Poems: 1946–1968* (Bloodaxe Books, 1986), reproduced with permission of Bloodaxe Books.

Every effort has been made to trace copyright holders of the extracts from poems quoted in this book. The author and publisher apologise if any material has been included without the appropriate permission and acknowledgement, and would be glad to be told of such omissions.

Contents

1 Introduction   1

Part I: Science and Philosophy (Finding Room to Breathe)   9

2 Light   11

3 The Structure of Science, Part 1   13
    3.1 A Case Study: Digital Computing   19
    3.2 Getting the Problem in View   21
    3.3 Symmetry in Physics   23
    3.4 Thermodynamics   30

4 The Structure of Science, Part 2   40
    4.1 The Embodiment Principle   40
    4.2 Biology   45
    4.3 The Role of Uncontrolled Change   50

5 Logic and Knowledge: The Babel Fallacy   54

6 Reflection   64

7 Purpose and Cause   74
    7.1 Why Has an Anteater Got a Long, Sticky Tongue?   80
    7.2 Science and Intellectual Discipline   90
    7.3 The Multi-Layered View   92

8 Darwinian Evolution   97

9 The Tree   113

Part II: Value and Meaning   119

10 What Science Can and Cannot Do   121
    10.1 A Brief Historical Survey   124
    10.2 Completeness and Cogency   133

11 What Must Be Embraced, Not Derived   141
    11.1 A Philosophical Investigation   143
    11.2 Reason and Faith   153

12 Religious Language	158
13 The Unframeable Picture	170
14 A Farewell to Hume	176
14.1 Introduction	177
14.2 The Argument from Lack of Explanatory Power	179
14.3 An Example Witness	185
14.4 Resolution: The Full Expression of Human Personhood	186
14.5 Four Witnesses	197
14.6 The Refutation of the Superfluity Argument	206
15 Drawing Threads Together	210
16 Extraterrestrial Life	216
17 Does the Universe Suggest Design, Purpose, Goodness, or Concern?	223

Part III: Breathing — 237

18 Silence	239
19 The Human Community	242
20 Encounter	253
21 The Human Being	255
22 Witnessed to	269
Appendix: Boyle's Law	273
Notes	279
Bibliography	281
Index	285

> ... the brightness over
> an interior horizon, which is science
> transfiguring itself in love's mirror.
> — the final phrase of the poem
> 'Destinations' by R. S. Thomas [70]

1

Introduction

> Six points to be shown in the book are introduced. Science is a rich tapestry which does not at all suggest that the world is a purposeless machine, but nor does it replace the arts and humanities. We respond to our situation through all our powers of expression, including forms of giving in which our very identity is shared and allowed to be shaped in return.

It is time to shake off a widely believed but mistaken idea of what science is and how it interacts with human life in the round.

For a long time now it has been the habit of science writers to present their discipline as if it was the be-all and end-all of knowledge, and everything else follows in its wake. Particle physicists have written about their forthcoming 'theory of everything' as if it amounted to the final word on the nature of reality, the very 'mind of God'. I accuse also the ten thousand websites, and the popular books that have spawned them, that have presented Darwinian evolution as if the authors knew the purpose of each organism's life to be the proliferation of that organism's genes. The same fundamental error is promoted by neuroscientists who, waxing lyrical over wonderful magnetic images of the living human brain, have declared or implied that all the functioning of the brain is about to be laid open, with no input from the arts and humanities required.

It is my opinion, and the central thesis of this book, that all such visionaries in fact have a skewed vision of both science and the arts. I think that what poets do, and what literary critics do, and what musicians do, is every bit as truthful and insightful about the nature of reality as is what physicists and chemists and biologists do (and mathematicians, engineers, historians, philosophers, and so on). Of course the afore-mentioned skewed visionaries will immediately claim that they agree, and they did not intend to deny or undermine the validity of the arts. But the fact is, the way they have presented science is such that science would and must do exactly that, whereas I think that when grasped correctly, it does not.

For thirty years I have had an intense desire to get a more truthful vision of the nature of the physical world and human life in it, a vision which does not fall into the error of misconceiving science and presenting it in hegemonic fashion, the way this is done in numerous popular books.

To be clear, let me say that I wholeheartedly welcome many of the points science writers make, especially about being honest and careful about evidence, and letting the world be what it is, and celebrating the richness that ordinary physical stuff is able to support, without overlaying it with superstition. Such writing gives voice to the widespread wish to affirm that the physical, material world is itself the expression of what matters, and it does matter, whether or not it is a temporary phenomenon. All this is welcome. But in our celebration of scientific knowledge we have to allow also for basic philosophical points, such as the difference between a text and its meaning. For example, when one human being meets another, they see plenty of evidence, but what conclusion do they form? The very same evidence might be interpreted to mean 'here is a slave that I can use' or 'here is fellow human being, every bit as worthy of consideration as myself'. It is not self-evident that people will always come to the second conclusion, and furthermore, it is not the case that any study of fundamental physics will suggest either that they should or should not. What shall we do, then? Is the 'modern, scientific' view the view that moral considerations are really just subtle power games with no intrinsic beauty and truth of their own? The point I want to argue in this book is that whereas the adjective 'modern' might apply to that conclusion, the adjective 'scientific' does not.

The overarching aim of this book is to get in view something that pays deep attention to the natural world and to the whole of what goes on in human life. I mean *attention* in the sense advocated by the French philosopher Simone Weil;[77] it is the decision to work hard at listening sympathetically and seriously to the realities that address us, both in the wider natural world and in our shared life. In the present context, it is the decision to listen sympathetically and seriously both to what science is and does, and to what other areas of discourse are and do. If one takes up this challenge, then one discovers something important and liberating. One discovers that the picture of the natural world that we obtain from scientific study is not the one widely assumed in contemporary culture. It is widely assumed that science is somehow 'on the side of', or implies, or is the natural partner to, the view that the physical cosmos, and life on Earth, is a sort of huge machine moving blindly into the future, and this motion has no purpose or meaning beyond the stories or meanings that we and other animals invent for ourselves. I aim to show that this is not so. Rather, scientific study, approached with a little bit of philosophical maturity, is perfectly in tune with, and indeed strongly suggests, the view that the natural world is capable of giving physical expression to a rich and deep range of languages, including all the ones that are important to human life, and in this rich expression we do not invent meaning but discover it. I mean here 'languages' such as justice, ethics, and aesthetics, to name a few. There is not just an appearance of these things in the dynamics of human affairs, but the reality of them.

When one states it bluntly like this, some may feel that this is obvious and does not need to be argued. However, it has repeatedly happened that scientists have supposed that their job is to reconfigure the whole human outlook, and replace all these languages by a single machine-like paradigm in which words like 'justice', 'good', and 'bad' have little or no meaning. It is important to show why such a supposition is wrong. Furthermore, this is not a case of a conflict in which the arty types fight a misconceived rearguard action against the inevitable truth of an outlook calling itself 'scientific', and which regards people as mere carrier bags for molecules. Rather, this latter picture has got science itself wrong, and is not genuinely 'scientific'.

What is going on here? How can an expert in physics, or in biology, or in neuroscience be at the same time wrong about physics, or biology, or neuroscience? It is partly owing to the compartmentalization of knowledge, and, I think, there is something of the myth of Pygmalion involved. In this story, the sculptor Pygmalion makes a statue of great beauty, and falls in love with it. The statue comes alive and various outcomes are imagined. In a similar way, a practitioner of any area of intellectual study may become so thrilled with the concepts at work in their area, that they fall in love with those concepts and begin to think they can explain everything. The concepts might be quantum fields and space-time, or genes and replication, or neural networks and synaptic potentials, for example. It seems so natural to think that each area can explain everything that is built out of the things studied in that area. Why not? What is wrong with this way of thinking?

The central argument of this book is twofold. I will argue, first, that scientific explanation does not take the mistaken form I just described, and, second, that analytical reasoning does not constitute the whole of what is required to get at truth. With regard to the first argument, I aim to show that scientific explanation is in fact two-way, not one-way, and that this naturally allows and indeed suggests the presence of further layers of meaning in the world, beyond the categories of science. Concerning the second, these further layers of meaning are explored by the arts and humanities, and, more humbly and tentatively, in spirituality. Those forms of spirituality which deserve attention are the forms which deal convincingly with the whole of this framework of meaning.

By saying that scientific explanation is two-way, not one-way, I mean that science is not a ladder in which physics explains chemistry which together explain biology, and so on. Rather, biology illuminates physics just as much as physics illuminates biology. The situation might be compared to the relationship between a stone and an arch (Figure 1.1). The nature of an arch is not *explained by* the nature of stones, even when

Figure 1.1. Arches and other architectural features at the Natural History Museum, London.
Source: David K. Hardman.

the arch is made of stones. After all, the arch could have been made of wood. Rather, the arch is an example of a shape which has certain generic properties, and such a shape can happen when stones are gathered and configured in a certain way. This is reasonably straightforward and obvious, but one must think it through at length in order to see the wide ramifications of this way of thinking for the whole scientific picture of the natural world.

Having got science right, we can venture into and respect the humanities, by which I mean the rich variety of further studies that people undertake in literature, arts, music, language, politics, and so on. The book does not address any specific issues within those subjects; it merely acknowledges their complete validity as disciplines having their own appropriate language and discourse. But having acknowledged all this, I want to go further, and ask what it is we ultimately rely on, and what we should aspire to. This is the question of how to help one another locate or 'frame' our lives in the most truthful and creative way.

It is widely agreed that the notions of *gratitude* and *generosity* are central to truthful living, and many (myself included) feel that *forgiveness* is equally central. In this book I am not going to argue about the relative merits of different values, except in a few general terms, but I will present some carefully constructed philosophical arguments in this area. The aim is to show the following points:

1. *The structure of the natural world is such that high-level principles describe and constrain what can be supported by low-level dynamics.*
2. A description of a physical object or process in terms of its internal structure does not address the question of whether or not the object or process serves a purpose in a wider context.
3. The question of whether the story of life on Earth has meaning beyond being a mere sequence of events ('just one thing after another') is not resolved by scientific study, but this does not imply that it is a non-question, nor that we have no means to come to reasonable judgements about that.
4. It is not possible to employ rational argument in order to derive one's most basic allegiance, or set of allegiances, from other considerations.
5. A certain well-known philosophical argument, found in various forms in Thomas Aquinas, David Hume, and Richard Dawkins, is unsound.
6. It is possible to give a coherent, meaningful, and attractive sense to the notion that we are called upon to respond to truth and

reality via the whole of our personhood, using all our powers of expression, including our very selves as personal beings.

The first item in this list is emphasized because it feeds into the discussion of the other items.

The argument referred to in item 5 is, in brief, the claim that theistic types of religious response are either superfluous or simply don't help in making sense of our experiences. Aquinas argued against this; Hume and Dawkins assert it.

In the last item I am alluding to the notion that truth is rich in nature, having a sufficient degree of richness that certain widely used religious metaphors are appropriate. That is to say, personal language such as 'Parent' is appropriate when approaching the universal reality that most deeply supports and emancipates all things. Here I am talking about that non-contingent reality correctly named 'God', but I avoided the word because it is so widely misused.

In a previous book I offered a perspective on science and religion[*] which was not so much an attempt to persuade as to display. I was inviting the reader into a gallery, to show some at least of what it is like to understand and accept science while also understanding and accepting what some forms of theistic religion are trying to say. The present book is more of an argument; it does set out a careful case for various positions. But I have limited aims, and they are mostly philosophical in nature, as indicated above.

In this book I am trying to convey an idea about the structure of science, and explanation more generally, that is sufficiently different from what people have been taught to assume that it cannot be neatly tacked on or fitted in to much contemporary discourse. Rather, that discourse itself must change. I will present arguments for this idea, but much of the work consists simply in getting the idea in view. This is like teaching a language or an artistic style: you have to begin to speak the language, or to allow the style to work on you, before you can assess its success. The book therefore includes some science, and some philosophy, and some poetry and a little polemic, and some other things.

---

[*] Religion (in its good forms) is an effort to recognize correctly what we truly depend on, and to seek what can properly be aspired to, and to respond appropriately to what we find. Bad religion is the abuse of this.

Part I of the book, Chapters 2–9, presents the structure of what happens in the various parts of scientific study, and hence what this tells us about the nature of the physical world. The scientific picture of the natural world is rich and fluid, moving between multiple types of discourse; the physical world is not a machine but a meeting place. In particular, Chapters 7 and 8 look at the Darwinian evolutionary process, with a view first to pointing out category errors in existing literature, and then to getting a sense of what kind of process it is.

Part II of the book, Chapters 10–17, looks at what science cannot do, and how discussion of values and of religious ideas operates. Science, though full of insight and usefulness in its own domain, remains severely limited in how it can help us. This is because science cannot answer the most basic, everyday questions we face, such as, 'What shall I do this afternoon?'. To answer such questions we must pay attention to further ways in which reality impinges upon us and addresses us. This is about the notion of *value*, and forming judgements about what would be *good* or *bad*, *better* or *worse*, and what is worthy of our commitment. The concept of *values* and the concept of *goodness* invoke from us a type of response which is not altogether the same as following a reasoned argument, and there is nothing irrational or unworthy about admitting this.

Part III, Chapters 18–22, presents, in a positive way, an outlook which tries to do justice to all the various themes. The aim is to show how certain mainstream religious ideas can inhabit this picture without awkwardness. The book is not shy of spirituality, and indeed it ventures to include a few spiritual exercises, but it holds back from presenting any fully defined framework of religious life. Rather, my aim is to present science in harmony with art in such a way that the journey to spirituality is available, not closed off. This is important because humans are spiritual creatures, and to refuse them this journey (for example by suggesting that one must part from reason in order to undertake it) is deeply unjust.

PART I
SCIENCE AND PHILOSOPHY (FINDING ROOM TO BREATHE)

2
Light

Once upon a time, there was light. Bright light. Not *a* light, but Light—everywhere and everywhen. Not just bright: brighter than bright.
—This bright?
—No, brighter.
—That bright?
—No! Brighter still! Dazzling! Burning! Dangerous!
—But what is light?
—Light is motion, light is energy, light is waves, light is streaming matter–antimatter-bombs of energy. Rushing too and fro, never ceasing, always moving as fast as possible. Never stopping, nor even slowing down.
—Is it all everywhere the same, this light?
—No. There are knots of light, and there are streamers of light, and flickering fibres of light. Like the Northern Lights! Only huger, and more intricate.
—Is it a mush? Is it a soup?
—No! There are patterns and pictures in the light. Look! Here is a knot where two flashes came together. One flash goes this way, one flash goes the other, and where they overlap an extra bump of brightness moves along the wave. Look! Now the bump has met another bump, and they circle around one another. They are weaving a helix.
—Oh! How pretty! How big are those bumps?
—They are tiny. Teeny tiny. But see how many of them there are! Look, whole crowds huddling together! Ten thousand billion billion here, one hundred thousand billion billion there. They make shapes.
—Yes! And I see the shapes moving around, bumping into one another. How can it be? Where is all this light coming from?
—Look some more. Can you see? Can you see?
—Yes! Yes! Here there is a cloud of light, and here there is a rock of light! And here there is watery light and here there is windy light! The

Science and Humanity. Andrew Steane, Oxford University Press (2018).
© Andrew Steane. DOI: 10.1093/oso/9780198824589.001.0001

wind is blowing and the water is tossing and turning. There are whole worlds of light! Oh, how can it be?

—Now look here: I want to show you a special place.

—Oh yes, I see huge globes have gathered together in the light, and light is streaming out from the largest of them, and is taken in by the others, and given back.

—Yes, and now see what patterns there are on the third globe. It is the greatest, most splendid, and also most painful, dance of light.

—I see it. I can hardly speak. It is breathtaking. It makes me want to weep.

—Are you weeping for joy or for sorrow?

—I don't know. Look! I have found a tiny pulsating globule in the light, oh, so beautiful! It is less than a hair's breadth of a hair's breadth across, but it is astonishing! It is a factory of light. It has little turning motors and little walking tweezers and intricate coiled-up coils of light inside it. And it is forever foraging and drinking and then it grows and splits. Oh, tell me how it can be! How can there be such things?

—Look, look some more, and then you will know.

—I see fish of light, and birds of light. I see trees of light, and animals of light. And now I see people of light. They are walking and meeting, and laughing and weeping. How fragile they are! Just a moment's break in the light, and they would vanish into nothingness!

—Do you see their thoughts?

—I see the expressions on their faces, and the movements of their bodies, and I see them speaking and listening to ripples they make in the light.

—And do you see them making things?

—Yes! They are making lots of things. They are making useful things and useless things. They are rearranging the light.

—But what are they really making?

—Oh! I see! They are making each other!

—Yes. Now where do you think this light is coming from?

3
The Structure of Science, Part 1

> The first major theme of the book is introduced. This is that science does not present a ladder or tower of explanation, but a network of mutually interacting and informing ideas. The digital computer illustrates the interplay of low-level and high-level language. The role of symmetry and *symmetry principles* in physics is discussed at length. I argue not just that symmetry is central to physics, but, more importantly, that it offers a subtle constraining influence that is not the same as cause and effect, but is nevertheless central to explanation and understanding. Numerous examples are invoked.

This chapter and the next will present the structure of scientific knowledge and explanation, and, at the same time, the structure of the natural world (since I think the former gives good insight into the latter). The present chapter approaches this by first surveying the territory and then focusing on what are called the physical sciences. The next chapter turns to the life sciences. I have avoided technical terms (whether scientific or philosophical) where possible. If they are used then they can help some readers but leave others less than confident of getting the details; I think nothing essential will be lost to the latter class of reader.

My first main point is fairly straightforward: it is that scientific knowledge is multi-layered, and this reflects a multi-layered nature of the natural world. For example, you can do good work in biology without having any knowledge of the quantum field theory which describes elementary particle interactions inside all the molecules that make up a living cell. This is because the functioning of a cell, or of an organism, already makes its own sense, without regard to the details of the fundamental physics (fascinating as those are) going on at the level of individual atomic nuclei. This idea of *making sense* is absolutely central to science, but the way it works is widely misunderstood in the modern world, and often misrepresented by popularizers of science.

Science and Humanity. Andrew Steane, Oxford University Press (2018).
© Andrew Steane. DOI: 10.1093/oso/9780198824589.001.0001

The next point is much more subtle. It is that this multi-layered structure of knowledge does not have the character of a vertical building (as the metaphor 'layer' suggests), but of a two-way network in which each node (i.e. each scientific discipline) illuminates its neighbours. It is not true that once one has understood the physics of individual atoms and molecules, for example, then the study of biology becomes redundant. It is true that living organisms are ensembles of atoms and molecules, and the evidence suggests that no other ingredient (any sort of 'life force' or 'essence') is involved in their make-up, but the ensemble itself is nevertheless a new kind of thing. A heap of apples is not just 'lots of apples'; it is a heap. It is different from a shower of apples, for example.[1] Similarly, moderately complex things such as large molecules, and very complex things such as living cells, require for their appropriate description and insightful understanding all sorts of concepts and languages which simply don't arise in the study of atoms individually.

The reader may feel ready to accept this general point, but fail to be aware of its full ramifications. One must think hard to grasp correctly what the ramifications are; this is the subject of the first six chapters of this book.

The scientific viewpoint is often portrayed as one that describes the natural world from the bottom up, as if the universe is a machine in which, once you have understood the component parts, you also have a complete understanding of the whole (see Figure 3.1*). That is what Stephen Hawking has implied in his popular books, for example (hence his decision not to delete that pithy, memorable, and greatly misleading, inept, and inapt phrase, 'for then we should know the mind of God', with which he rounded off his *Brief History of Time*). Richard Dawkins 'doth provoke and unprovoke'[61] a similar view. He waxes lyrical about the explanatory capability of the mechanisms of genetic inheritance, variation, and natural selection, as if all the behaviours of living things should and must be seen as just so many robots or bags for replicator molecules. But one might ask, why pick genes as the basic concept, why not the laws of thermodynamics, or the quantum matter-wave fields that Prof Hawking has so expertly studied? Equally, if one

---

* Figures 3.1 and 3.2 show just a few broad areas, rather than the large patchwork of scientific disciplines, not to suggest that this patchwork falls neatly into compartments (it does not) but in order to bring out the issue under discussion in the text without further complicating the diagram.

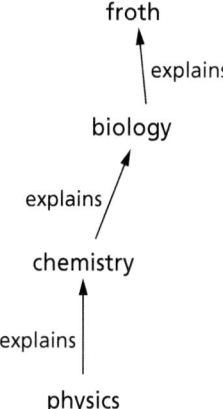

Figure 3.1. A WRONG view of the structure of scientific explanation. I cannot overstate how wrong this is.

goes as far as genes rather than thermodynamics, why stop there? What about all the further patterns which arch over these, and which form the structure of ethics and of human society? I will argue later that Dawkins's work, for all its admitted strengths, is deeply flawed by its failure to grasp the structure of scientific explanation correctly. This is not to say that scientists in general are unaware of this issue; it is to say that popular science writing has not handled it well.

I invoked a couple of famous names in the previous paragraph in order to make my point that the structure of scientific explanation is misrepresented in widely read books. I don't for a moment think that the two people I have mentioned would be incapable of appreciating the merit of other areas of scientific and human understanding; far from it. It is just that their written work sells the 'bottom-up' picture of science, as if that is the truth of how scientific explanation works.

It is not.

The truth is much more rich and interesting than this. Again and again in science we find that it is only by first understanding the whole that one learns what the ensemble of the component parts is capable of doing. The Nobel prize for biology is regularly given to people who are totally ignorant of Wick's theorem, for example, even though Wick's theorem (a theorem in quantum physics) plays a role in every elementary particle interaction in every molecule in every living cell.

Similarly, the Large Hadron Collider at CERN, Geneva, tells us some beautiful and fascinating truths about physics, but close to nothing about endocrinology.

Many an expert biologist may never have even heard of Wick's theorem, but this does not diminish their grasp of the area of science they work in, nor the value and validity of their contribution. Similarly, if some genius did manage to discover a fundamental physical theory that left nothing out and was precise in every detail, then I would heap Nobel prizes upon her, but I would not for that reason think she had escaped the same follies and misperceptions of human relations and values that beset the rest of us. All this is because the world is structured a certain way, in which different layers of complex behaviour already make sense in their own terms, irrespective of much of what goes on in other layers. Each also illuminates the understanding of other layers, and *this illumination proceeds in both directions*. Figure 3.2 is an attempt to indicate this structure.

When some popularizers of physics write about a 'theory of everything', in reference to a collection of quantum field theories, and imply that this provides a thorough description of the universe and everything in it, they are not being nearly as scientific as they suppose. They are in fact being, to a significant extent, misleading, because in fact to possess nothing but a complete and correct account of fundamental physics (if we suppose for a moment that that were possible) is to be in almost complete ignorance of the nature of the physical world and what goes on in it. One's insight would be comparable to that of someone who has learned the rules of chess but never played the game. All the insight that grand masters take years to acquire would be lacking.

I myself know the rules of chess, but I don't in the least think that this gives me an insight and understanding of chess remotely comparable to that of a great player such as Garry Kasparov. Of course not. This is too obvious.*

---

* In any case, a so-called theory of everything is itself questionable. How could it be shown to be correct? If it is really a theory of *everything*, one which somehow captures everything that goes on in human brains when we are checking it, then on what basis is the check carried out? This whole area gets caught up in the puzzles and paradoxes of self-referential assertions, such as the statement, 'this statement is true'. This seemingly innocuous statement may well be true, but that is not the only possibility. It might equally be false, and then what it asserts would indeed be false. So one can assign to it either truth value. An even more puzzling statement is Gödel's one: 'this statement is false'. Now it cannot be assigned a truth value. One can only make sense of it by

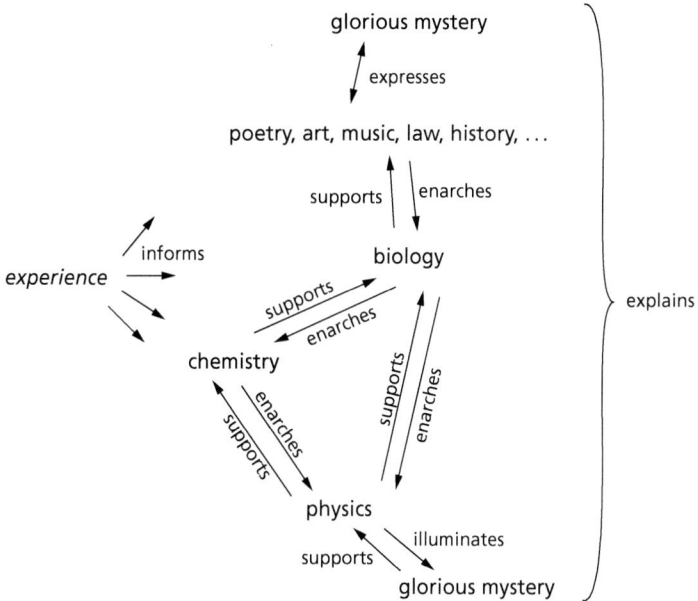

Figure 3.2. An attempt to indicate correctly the structure of scientific explanation and the way things are. The word 'supports' here is a shorthand for 'allows and physically embodies the expression of'. The term 'enarches' stands for 'exhibits the structures and behaviours that make sense in their own terms and are possible within the framework of'. The word 'mystery' refers to that which is not fully comprehensible by ourselves; this does not mean utterly incomprehensible, nor that one should give up the attempt to comprehend better. I call it 'glorious' in order to say that this is not a negative. It is a good thing that we don't know fully and cannot grasp completely our own existence. Also, the fact of that existence is gloriously positive, especially in view of its very great richness of expression.

In the case of chess, it is indeed obvious that the rules don't illuminate the game very much, but one might object that this is an unfair comparison because human or other agents are needed to choose the moves in chess, whereas the 'game' of the universe plays itself. It is this objection that requires a careful response, and that is why there is a

expanding it, requiring further explanations of the language that has been adopted and hence the context in which truth claims are being made. This is a footnote in the present chapter because I am presenting other things, but it is a useful reminder of the subtlety of logic and the limited nature of all knowledge.

lengthy discussion in this and the next chapter. In Figure 3.2 I have coined the term 'enarch' to state briefly the relation of a subject such as biology to physics; it is the idea that the structures on show and at work in biology relate to the principles of physics in somewhat the same way as arches do to stones. They are not determined inexorably by microscopic laws, nor the result of pure chance. They are truths that the processes of the world are able to discover and exhibit.

A book that aims to be thoughtful about these issues is Prof. Sean Carroll's recent work, *The Big Picture*.[19] It is wide-ranging and has the intention of being fair to subjects outside of physics, but in the end, and repeatedly, Carroll's work makes the fundamental error that I am pointing out. Carroll wants to admit the validity and value of art and literature and social justice and moral philosophy, but when push comes to shove, he sweeps it all away. Fundamental physics wins the day. 'Life is short, and certainly never happens'; 'It's just the quantum wave function', he writes. 'Everything else is a convenient way of talking.' But by what argument is this conclusion arrived at? By a faulty argument. That is what I want to show in this book.

A similar mistake to the physicists' 'theory of everything' is made when people working in biology or zoology claim or imply that the process of genetic inheritance and Darwinian evolution gives a biological 'theory of everything', as if there is nothing further to say. In both cases I do not intend to detract from the intellectual value, and indeed glorious insight, of working principles of the widest possible reach. But in both cases the structure and behaviour that can be supported by the underlying dynamics are not subsumed by the rules of those underlying dynamics. I will be expounding this at length in this and the following chapters. It is a fact that is especially relevant in Darwinian evolution, because that is a process of exploration. It has been the gradual process of development of every living thing, but an exploration process cannot dictate what will be found. It can only search. What is found is determined by other considerations, such as the consideration of what is physically possible, and of what ways of living can function and prosper. The significant point here is that what is found is not entirely random, and is only partially explained by the search process. This is what Richard Dawkins has failed to appreciate in the central argument of his book, *The Blind Watchmaker*, when he seeks to argue that the Darwinian process amounts to 'the only known theory that could, in principle, solve the mystery of our existence.' What he means is that the only satisfactory account of some highly structured physical object is one in which that object is shown to have come about by the operation of an undirected combination of less highly structured things. This contains a truth, but it is not the whole truth, because it fails to appreciate that the very structure arrived at is structure, as opposed to random shaping and flapping, only because it embodies the patterns that succeed in the area of the physical world in which it operates. Those patterns are themselves an important part of reality, and they are not explained by the Darwinian process, just as the laws of arithmetic are not. Rather, they constrain what evolution can discover. This is an important point and I will return to it.

Now, if scientific knowledge and the world it seeks to grasp are not structured in the simplistic way that I am rejecting, then how are they structured? I have already briefly sketched my answer to this question, which is illustrated in Figure 3.2. It will take a considerable amount of careful argument to clarify what this diagram is intended to represent, and to show that what it presents is a correct picture of the structure of scientic explanation, or at least a much closer approximation to the truth than what is indicated in Figure 3.1.

To understand the relationships indicated in Figure 3.2, it will be very helpful to consider the role of *symmetry principles* in fundamental physics. But before I present that, it will first be helpful to get at the concept of different languages or levels of discourse, and the example of digital computing provides a convenient way to do this.

3.1 A Case Study: Digital Computing

Consider the functioning of an ordinary digital computer, of the type based on binary logic and widely used in the modern world. One can describe what happens in such a machine in terms of electrons and conduction, or in terms of circuit theory such as transistor action, or in terms of binary logic gates, or in terms of very primitive programming called assembler code,* or in terms of overall machine architecture and a high-level programming language. I assert that the high-level programming language is *explanatory* in this example, in the sense that

---

* Assembler code consists of long lists of very simple instructions, such as 'move the contents of memory location A to processor location R'; a high-level language deals in larger objects such as a video image or a database and contains broader instructions such as 'gather all the items whose image resembles the first image'.

it aids understanding and is a legitimate and indeed important part of the complete picture.

I think this assertion is so natural that if one wished to deny it then the burden of proof would lie with one who so wished. However, the following feature is noteworthy.

Consider, for the sake of argument, a deterministic machine (that is, one with no random element in its functioning), and suppose the computer hardware is fixed, and a stored program is about to run, as soon as a key is pressed. In this case if one only possessed the logic-gate description, then this would suffice to allow one to perfectly predict the evolution over time of that set of logic gates and memory elements. No description or knowledge or understanding of the program at a higher level would be needed. The description would start out with a list of binary values at all the memory locations before the program started, and it would *correctly* predict the binary values at *all* the memory locations when the program finished. So would it be legitimate for someone owning such a description to judge that they had got a complete explanatory account of the computer, even though they were utterly ignorant of the very notion of a high-level programming language? The answer, I assert, is 'no'. This is because the logic-gate description, even though it has perfect predictive power in terms of what the logic gates do, has no predictive power in another sense—the sense of what the machinations amount to; what computational task is being achieved. It cannot predict that because it does not even 'see' that. The low-level description fails to capture what the computer is, because it does not even begin to express most of the significant facts about what a computer does.

The higher-level language in this case is the language of *assignment statements* and *if-then-else clauses* and *iterative loops* and *variables, functions, programs*, etc. This is a language or set of concepts that makes sense in its own terms. Also, there is a one-to-many relationship between the high-level process and the types of microscopic hardware that can support it. That is, the same program can be run on a number of different machines, with not just different hardware but different *types* of hardware, and different operating systems.

It is untrue to say that a language such as Java or C or Matlab is merely an outworking of logic-gate behaviours, because one does not need to know about such behaviours in order to understand a program, or simulate it on another machine, or follow its instructions by laborious pencil-and-paper calculations. Nor is a high-level language merely a shorthand for logic-gate behaviours, because a given high-level behaviour can be supported by a variety of low-level hardwares, and one can also see the relationship the other way round, where the programming language is designed first and then the logic gates are so arranged as to support the language and respect its protocols. Indeed, computing devices are mostly designed this way. Furthermore, programmers do not need to know much about computer hardware. In practice most programmers know almost nothing at all about the levels lower still, such as solid state physics and quantum electrodynamics.

For all the reasons outlined in this section, the explanation of the workings of a computer offered by a purely logic-gate description is such a blind and dumb (i.e. unable to see or speak) sort of explanation as to amount to almost no explanation at all.

3.2 Getting the Problem in View

The points made above about digital computing are reasonably easy to make and to grasp, because there we already know at the outset what sorts of higher-level language are going on, and also because we know that the computer was designed expressly to support those types of languages. When it comes to the natural world, we have to proceed much more circumspectly, because we must not claim to know things we do not know, such as what types of concept and language apply at the highest level; and nor is it at all easy to make sense of the idea that the whole natural world was created to some purpose by one who could have intention. Although, later in the book, I am going to offer some material which, tentatively and incompletely, opens out what it might mean to speak like that, I am not going to assume it here.

If we don't claim to know what the world is for, nor what are the richest meanings to be found in it, then we can't make the analogy to the digital computer quickly or directly. Nevertheless, the different language levels illustrated by the computer act as a useful prompt. They alert us to the idea that an 'explanation' in terms of one set of concepts does not necessarily replace or make redundant an explanation in terms of another. Note, my point is not that the universe is like a vast digital computer (I think it is in some respects like that, but very likely this will turn out to be an inadequate model, and so will the model offered by quantum computing). I intend by this computer example only to illustrate a point about languages and levels of explanation. It also illustrates the important point that the higher-level descriptions are not the 'junior partner', mere afterthoughts which get tacked on, but which are not really necessary. That is completely wrong. It is close to the opposite of the truth.

However, as I say, setting out the application of this sort of idea to the processes of the natural world is not straightforward. We have become very familiar with the notion 'just get to work on the bits and pieces, and the whole will follow', so that we hardly know how to put it another way. This is because of a widespread failure of science education, which I am trying to remedy by this book. I am by no means the first to insist that scientific explanation does not have the structure of a ladder, but it is a truth that has not yet been widely seen and appreciated. People working in the humanities might find it rather obvious, but it has become painfully and disastrously common for scientists to deny the very meaning of discourse that they (wrongly) think their discipline renders redundant. An example might be for one to study the human brain at length, and then announce, in the name of science, that concepts such as mercy and forgiveness are meaningless. The illogic of such a conclusion seems to many of us obvious, but it is becoming commonplace to let it pass. This particular example is drawn from an interview on BBC Radio 4 in the UK, and the interviewer did not have the confidence to question the pronouncement, because it came from a neuroscientist whose work *in neuroscience* deserved respect.

The reason why a capable scientist might assert something nonsensical (assert in words, that is; I doubt they would assert it by the way they live), such as that forgiveness is without meaning, is because they have bought into the idea that brains are machines and machines can neither need nor offer mercy. I would reply that brains are machines, if by a machine you mean a physical entity with moving parts and many regular patterns of motion; but I do not for a moment accept that such a physical entity cannot exhibit such qualities as mercy, because the patterns might be of a type (not yet understood by us) that allows this. I could agree that the limited grasp we have of the brain might be such that current models do not describe something which supports the expression of genuine mercy. I do not agree that there can be no model which is both correct at the molecular level and also admits the validity of the high-level language commonly adopted in human relations. In other words, if we allow that the noun 'machine' applies to the brain, then we are using the word 'machine' in a broad way which includes types of functioning that are not yet well understood. If we adopt that usage, then the brain is *both* an entropy-generating machine *and* an empathy-generating machine, and the second fact is as genuine as the first.

Now to return to the job in hand. We need to explore how it can be that different areas of scientific discourse do not replace one another but still connect with one another. This is not just a question about the human activity called 'scientific discourse'. It is a question about the nature of the physical world, and it already arises in the area of science which is my own area of expertise, namely fundamental physics.

3.3 Symmetry in Physics

The notion of 'symmetry' is very important in physics. The word is used in a certain technical sense, one that includes the notion of mirror symmetry but which also extends to other, similar, ideas. An object, such as a crystalline solid, which would not change if it underwent a mirror reflection, is said to posses 'mirror symmetry' or 'reflection symmetry'. An object, such as a circle, which does not change when it is rotated about a suitable axis, is said to possess 'rotational symmetry'.

One can also have symmetries of more abstract things, such as equations and laws of motion. An equation that describes particle collisions, for example, might have some form of symmetry. If it doesn't matter if you reflected the particle trajectories (i.e. if the equation comes out the same), for example, then we say the equation has mirror reflection symmetry. If the equation does not need to mention the location of other things not involved in the collision, then we say the equation has 'translational symmetry'. The term indicates that it would not matter if one moved or translated all the particles involved in the collision from one place to another, by the same displacement: the equation remains the same, and therefore so does the behaviour it describes.

A *symmetry principle* is a principle that says certain types of change should not make any difference to what is going on in some physical behaviour. For example, a car engine functions the same way, no matter where the car happens to be. If you drive around Paris, or China for that matter, it does not matter how far you are from the Eiffel Tower, as far as the engine is concerned. All that matters to the functioning of the car engine is the local gradient of the road and the local temperature and air quality, and things like that. The position of other unconnected objects such as towers and trees does not make any difference. This is an example of the symmetry principle called 'translational invariance'. The workings of the engine do not change—they are *invariant*—when the location of the car is changed or 'translated' from one place to another.

The principle of translational invariance is a very important part of the principles of physics, and when brought to bear with the power of mathematics it yields rich and powerful insights into many physical phenomena. It also illustrates the structure of scientific knowledge in an important way that I will now try to bring out.

Suppose an engineer looks into the functioning of a car engine. He or she will use, for example, Newton's laws of motion (such as the one that says the acceleration is proportional to the force) and various laws of chemistry describing the burning of petrol. The engineer might never make explicit use of the principle of translational invariance. So we can imagine the engineer taking the following point of view:

> Look, here is the equation of motion: the force causes the piston head to accelerate, and the force is given by the formula for pressure in a hot gas, expressed in any convenient notation. That's it. That's all we need. All we need to do is find the solution, for given starting conditions, and we have everything. You see, that symmetry principle you mentioned—translation symmetry, or whatever it was—that is completely irrelevant. It has no role at all. It does not add anything. I have no need of that hypothesis.

This is the simple yet subtle point. What is the role of the symmetry principle here? If we have the formula that describes the force on the parts of the engine, and the equation that describes their motion in response to this force, then the symmetry principle seems to be an afterthought, and we can manage perfectly well without it. Or so someone might say. What should we think about that? Now our hypothetical engineer asserts:

> The equation of motion captures the sequence of cause and effect and correctly predicts or describes the behaviour. The symmetry principle makes no difference whatsoever to this. It adds nothing to the predictive power. The symmetry, or the lack of it, is simply a consequence, a property that the equation of motion happens to have. It could have turned out another way, but it happened to turn out this way.

What this reaction has failed to grasp is that the symmetry principle already makes its contribution before we ever write or discover the formulas and equations, *because it places conditions on what sorts of equations could make sense*. And science is all about making sense, or finding the sense that can be made. Symmetry principles in fact play an important role, because they amount to meta-laws which express higher-level principles that basic laws of motion must respect if they are to make certain types of sense. Such principles are of extreme practical importance in physics. They have been central to making progress in fundamental physics for over a century. But what is at the heart of this is not merely that symmetry is helpful to us (it helps us to formulate correct physical models), but also that symmetry gives *deeper insight into what we find*.

The example of the car and translational invariance also illustrates a further important idea that illuminates the whole of science. This is that *the truths of the whole are not negated by the truths of the parts*. The truth of the whole in this example is the simple but important observation that the functioning of a car engine does not depend on the location of the car. The truth of the parts is the behaviour of the pistons and fuel and the equations that describe them. The physics and chemistry of those motions have translational invariance, and this is an important, insightful, and simplifying observation that gets to grips with the big picture without needing to trouble about the details. This truth about the whole is not negated by the truths about the parts.

The example also goes further. Suppose someone says, 'Thanks for pointing out that symmetry to me. I see it is right. It is an overall property that follows from the equations describing engines. But still, those equations are the main thing. The translational symmetry is interesting, but it might have come out another way.' The further point here is that there is a sense in which the symmetry is prior, and unavoidable. As soon as we focus our attention on the car engine, and consider that it can be described in and of itself (with just some input from its immediate environment), then we have *already* assumed the symmetry called translational invariance. This is because, if the symmetry were not there, then the engine would depend on the location of other things after all, and therefore could not be described only in terms of itself and its immediate environment. So the claim that the symmetry 'just happens' to emerge from the equations of motion of the car engine is wrong. As soon as you even suppose that there is an equation for the car engine, it must have this symmetry.

It follows that the overarching fact captured in the symmetry principle is not an effect coming from a microscopic cause. Nor is the symmetry merely a pleasing or helpful thought. Rather, the symmetry principle acts in a subtle but powerful way, constraining the very equations that we find in fundamental physics.

Consider next a richer type of symmetry that is well known to physicists, called the principle of relativity. This is the principle that asserts that constant-velocity motion can only be defined in a relative sense. If you sit on a smoothly running train, then you cannot tell whether the train is proceeding on its journey unless you look out of the window. This is because observations inside the train, such as experimenting with objects rolling on the floor, or shining light-beams around, will give the same results when the train is standing at a station as when the train is moving at constant velocity on a smooth track. Similarly, no experiments performed in the confines of your living room can tell you the speed of your living room relative to another object, such as a bus on the high street, or one of the comets orbiting the sun. This principle of relativity, like the principle of translational invariance, also yields rich insights into physics. When combined with another statement called the 'light speed postulate' it enables us to predict the effects of motion on the shape of solid objects. In particular, by carefully considering what is meant by words such as 'simultaneous' and 'distance', we come to expect a certain distortion in the shape of fast-moving objects, called 'Lorentz contraction'.

Now there is an interesting bit of history here, which powerfully illustrates the point of this chapter. This shape-change of solid objects, 'Lorentz contraction', was first predicted by George FitzGerald, Hendrik Lorentz, and Joseph Larmor at the end of the nineteenth century on the basis of the theory of electric and magnetic fields. By using a set of eight differential equations, some hypotheses about the internal structure of crystalline solids, and a lengthy calculation, one can show that the Lorentz contraction must happen if the fields inside material bodies obey the equations of electricity and magnetism. But the other method of calculation—the symmetry method—is much more general and much more direct. It gets to the heart of the problem in a few steps of piercing insight. This was the contribution of Einstein's great paper of 1905.[28]

This is an example where two different methods of calculation provide the same conclusion. On one hand there is the lengthy analysis from the equations of electromagnetism, which are known as Maxwell's equations. On the other hand there is the much more general and direct method of grasping what the symmetry principles are, and realising what they imply.

But, like the car engine example, someone might want to deny that the symmetry method is really adding anything. 'The principle of relativity is all very well,' they might say, 'but it does not constitute an explanation. The *cause* here is the behaviour of the electromagnetic field. The symmetry principles are just ways of referring to properties that the electromagnetic field happens to have.'

Someone taking the above point of view sees the world as all about cause and effect, and concepts such as symmetry principles do not seem to fit into such a scheme in a cause-like role, so they are brought in simply as interesting observations, a sort of colouring-in, an embellishment that we could, if we like, do without. Someone might say that these symmetry principles are just irrelevant add-ons which bring nothing to the party. They might say they are 'not explanatory'. They might judge that progress in science involves discovering equations and analyses which liberate us from outdated ways of thinking—non-analytical ways which add nothing, just as the principle of relativity adds nothing to electromagnetism.

But of course it adds nothing, I reply. *It adds nothing because it has already added everything.*

In fact, the equations of electromagnetism can themselves be *derived from* the symmetry principles and a few general notions of what makes a simple field theory. I don't expect the reader to be expert enough to assess this specific example. Most readers will have to take that final statement on trust, but it is not in any sort of dispute. It is part of the textbook knowledge of this area of science.[56] My purpose here is to use it as an illustration of something that is all-pervasive in science, namely this: *ideas such as symmetry, which express truths about physical behaviour without needing to specify the microscopic mechanisms, are not the result of microscopic processes. Rather, they are principles that microscopic processes cannot help but respect if they are to make sense in certain specific ways.*

And this runs deep. The concept of translational symmetry, for example, is part of the very meaning of the word 'isolated', as I already mentioned. If a set of entities is isolated then its behaviour logically *cannot* depend on the location of other entities because if it did then the set would not be isolated after all. The realization of this is a profound step forward, but it is not a discovery of any particular equation. It is a realization of what types of equation make sense, or can support a larger type of sense. As I already commented, the symmetry principle both helps us to find the right equation, and, crucially, it shows us *what sort of equation we have*. The symmetry principle is first a guide, and then, in a certain hard-to-express but beautiful sense, it 'inhabits' the equations of physics. The concrete phenomena that are in the world are a sort of physical embodiment of the symmetry principles.

By moulding our mathematical notation, such insights shape the very way we 'see' the world. We adopt a particular mathematical notation called *vector notation* when dealing with classical forces, and we adopt *tensor notation* in relativity. This shapes not only what equations we find, but also what equations we even put forward for consideration. This is very important in helping us navigate the narrow thread of ideas that make sense, amongst the great morass of ideas that do not. Much of the progress in fundamental physics in the twentieth century can be seen as a sequence of triumphs of reasoning from symmetry.

Furthermore, these insights survived the huge transition from classical to quantum physics that took place in the twentieth century. That is a very striking fact. It illustrates that the symmetry principles have a validity in their own right, independent of the underlying language (that is, classical properties or quantum operators) in which they are expressed. As Philip Anderson (Nobel Prize for Physics, 1977) put it in a famous paper,[4]

> It is only slightly overstating the case to say that physics is the study of symmetry.

3.3.1 Avoiding an Overstatement

Having emphasized the role of symmetry in physics, it behoves me to remind the reader that it is not the whole story. The above statement by Anderson is, as he says, an overstatement. I don't want to suggest that one could correctly guess the symmetries of the natural world without experimental observation, or that one could correctly construct basic physics from symmetry alone, because such suggestions would be false. Indeed, there are examples of symmetries which were widely expected but not found, or found in a qualified sense. An example is mirror-reflection symmetry, or handedness, which I will briefly discuss next. Nevertheless, the role of symmetry remains as I have discussed: central, relevant, and in a subtle relationship with the equations of physics.

Mirror-reflection symmetry merits a further comment. The physicists' technical term for this type of symmetry, when applied to abstract things such as equations and types of process, is *parity*, but we won't need that term. The notion is readily illustrated by some examples.

Cars intended for sale in the United Kingdom are designed as 'right-hand drive'. That is, the steering wheel and other controls are on the right-hand side of the car (when one faces towards the front of the car). However, one can equally well make a car for sale in another country, with the driving controls on the left. The second car will work equally well. This is because the physical processes, such as burning of petrol, electrical signals, and mechanical transmission, do not have any built-in preference for one handedness over the other.

This type of symmetry is very widespread in nature. When one investigates almost any process going on in the natural world, one can imagine holding up a mirror to the scene in question, and the situation and processes observed in the mirror will be a state of affairs that could possibly come about somewhere. As far as we know, this is true for all processes based on gravity, electromagnetism, and strong nuclear forces. However, it turns out that certain types of radioactive decay do not exhibit such symmetry. Sometimes a perfectly symmetric sphere such as an atomic nucleus will emit electrons in a preferential direction relative to the sense of the intrinsic angular momentum (loosely speaking, the rotation) of the sphere. This is highly surprising. To get the sense of surprise and interest experienced by physicists when this was discovered, compare it to the following analogy.

Suppose there were a factory for corkscrews, and the directors of the manufacturing company wanted to make available to the public corkscrews in one of two colours, red or blue, and also of either handedness, right or left. They duly set up the factory, and it is found that all the right-handed corkscrews come out coloured red, and all the left-handed ones are coloured blue. But the directors don't want that; they want both types of corkscrew to be available in either colour. So they instruct the engineers to sort it out. The engineers duly investigate the manufacturing process, and they discover, to their surprise, that the connection between handedness and colour is not owing to any design error in any of the machines they have made, but is forced on them by the very nature of the physical world. In fact this would not happen for corkscrews, but it is an analogy for what has been discovered about fundamental particles called neutrinos. One can make corkscrews of either handedness, but neutrinos come in only one handedness, and anti-neutrinos the other.

The radioactive processes involving neutrinos are said to 'break' or 'not conserve' mirror-reflection symmetry, so here we have an example of a symmetry which was thought would be respected in all natural processes, but is not. Does this then undermine what I have been saying about symmetry? Arguably, it does not, because the breaking of mirror-reflection symmetry in this example is not arbitrary. It is itself elegantly accounted for as part of a larger mathematical framework that is constrained by the demands of the symmetry called Lorentz covariance, which is the one that is associated with the principle of relativity. When one asks what kind of mathematical structures most naturally encode or exhibit Lorentz covariance, one finds that they include mathematical objects called *spinors*, and these spinors exhibit the type of handedness which is seen in the physical things called neutrinos. In the end, this example tends more to reinforce Anderson's summary statement than to undermine it. But it remains true to say, as I already have, that reasoning from symmetry is not the whole of what is required in order to do physics.

3.4 Thermodynamics

Let us now explore the sense in which symmetry-like reasoning is involved in the overlap between physics and chemistry.

Modern physics may be regarded as the study of the interplay and outworking of three main notions. These are first the notion of space-time and its geometry, which is the subject of general relativity; second the notion of quantum amplitudes and quantum operators, which are basic components of quantum mechanics; and third the notions of entropy and thermal behaviour, which are the subject of thermodynamics.

General relativity has as its foundational idea the symmetry which gives the subject its name; it is a symmetry which captures the idea that physical behaviour in space and time ought not to depend on how one chooses to map space and time. Quantum mechanics is a general mathematical method in which symmetries of many different kinds can be expressed with economy, and these together go to form what is called the Standard Model of particle physics. This model contains most of what we know about fundamental particles and fields.

Thermodynamics introduces some of the basic tools that one needs in order to think about composite systems with very many parts and in which there is a certain amount of random fluctuation going on.

The main principles of thermodynamics, called the first and second laws of thermodynamics, are themselves symmetry principles (to be precise, the first follows from a symmetry principle, and the second asserts a certain specific non-symmetry). These laws make generic assertions about what sorts of behaviour can make the kinds of sense that are expressed in the physical world, without worrying about the microscopic details. Together they elucidate the general conditions in which there can be large-scale behaviour that has the combination of continuity (the first law) and development (the second law). Much of chemistry, materials science, engineering, and biology is elucidated by these laws.

Now let us consider a commonly made assertion. It is commonly said that chemistry is 'explained by' physics. This is an example of the 'ladder of explanation' picture of science: science as a kind of high-rise tower, with physics on the ground floor and chemistry on the floor above. One can understand why this is said, up to a point. Quantum mechanics gives a beautiful mathematical apparatus which, when applied to electrons and protons, correctly predicts important properties of the chemical elements, such as chemical valence and types of chemical bond, and these properties would otherwise be mysterious. Therefore there is indeed a sense in which physics is highly explanatory of chemistry. However, we (scientists in general) have found it useful to retain the term 'chemistry' because much of chemistry concerns the study of more complex things, such as large molecules, and chemical reaction networks, and in these studies the properties that are elucidated directly by physics are not necessarily the terms of discourse. One begins to find it useful, and indeed necessary, if one wishes to gain understanding, to adopt a higher-level language. It is like the step from logic gates to programming languages in the computer. One cannot manage without the logic gates, but they are of very little use in understanding the programming language. Chemistry in fact adopts a toolbox of different ways to get insight into molecules, most of them only loosely related to quantum physics. If follows that it is at best a half-truth, and at worst just plain wrong, to say that physics explains chemistry. Physics no more explains chemistry than logic gates explain programming languages.

The change of focus that takes place between physics and chemistry is not clear-cut, however, and many of the concepts of particle physics retain their usefulness in chemistry. So the analogy with the digital computer is not complete. Nevertheless, that analogy is relevant. This is why undergraduate chemistry courses do not need to include any instruction on some parts of quantum field theory that physics undergraduates have to master.

The sense in which chemistry is supported, but not explained, by particle physics is well brought out by the relationship between thermodynamics and particle physics.

The first and second laws of thermodynamics provide basic components of the scientific language of thermodynamics. They give defining statements about energy and entropy,* and from them other things can be derived, such as the concept of temperature and how it relates to heat transfer. A set of relations called the *Maxwell relations* can be derived, and they show how various physical phenomena relate to one another. For example, they can be used: to find how the pressure-dependence of the boiling point of a liquid relates to the change in volume and the energy supplied; to find relationships between the compressibility of a physical object and its capacity to take in heat; to extract some basic information about the magnetic behaviour of a superconducting metal; and to find a large variety of further relationships.

But all this is possible without ever needing to write down, bring in, or have any knowledge at all of quantum mechanics or particle physics!

Now, it is true that all these thermodynamic relations are in practice expressed through the behaviour of physical entities which are in turn correctly described by quantum mechanics and particle physics. Also, there exist arguments to show that the laws of thermodynamics must themselves follow if the equations of particle physics are as we think they are.† The point is, however, that this does not necessarily imply that thermodynamics is 'explained by' particle physics, because the

---

* I will be using the word *entropy* in its strict definition as a technical term within physics, but the reader need not be familiar with the quantitative details. It suffices to note that entropy plays a large role in understanding change, especially the direction of change, in many physical processes. Entropy is a physical property which is transferred during heat flow and which cannot decrease in an isolated system; in a specific technical sense the entropy of a system or body quantifies the absence of structure in that system or body.

† Actually, there is a persistent difficulty in getting this fully proven, but that technical issue is not my concern here.

Figure 3.3. Georges Seurat, *Embroidery, The Artist's Mother*. Detail, conté crayon. *Source*: The Met, New York; Creative Commons.

relationship has essentially the character of a symmetry principle. One may, with equal claim on our attention, assert that particle physics is shaped by thermodynamics. That is, one may assert that, whatever the equations of particle physics may be, they must respect energy conservation and they must respect the nature of entropy.

In the end, science itself cannot adjudicate which if any of these ways of seeing the relationship is better. However, we can say, with certainty, that a higher-level language is not rendered irrelevant by the mere fact that it is supported on a lower-level structure (the digital computer illustrated this). Similarly, a pointillist artwork does not fail to show an image (Figure 3.3). The patterns traced by thermodynamics are no less real and significant than the ones traced by the equations of particle physics. Overall, the reaction which makes most sense, and which best captures the notion of 'explanation', is one which embraces *both* areas of discourse and refuses to allow one to displace the other.

What this means is that the laws of thermodynamics not only underpin much of chemistry, but also place constraints on microphysics (quantum theory, etc.). This constraining influence is not a cause—effect relationship (quantum physics is not an effect coming from a thermodynamic cause), but it is a symmetry-like relationship, or a form of logical relationship. It says that if the universe is to have some sort of large-scale sense, with developing pattern, then the microscopic laws of motion can only be of a certain kind. It is like saying that if the game of chess is to be engaging, then the basic rules must be neither too simple nor too arbitrary.

This way of 'reading' thermodynamics has not been widely taught in physics courses in recent years; it is probably fair to say that in the current climate it would be controversial. However, it is not at all controversial that there exists the important interplay between symmetries and the equations of basic particle physics that we explored in Section 3.3. It is also extremely hard to argue that the second law of thermodynamics is a less reliable guide to the nature of the physical world than is our current best bet at the equations of particle physics. It follows from this that the idea expressed in the previous paragraph has very strong credentials.

So let me repeat it.

The point I am making is that, at the level of the most basic properties of things existing and moving in space and time, the 'ladder' picture of science is already wrong. In the study of particles and waves and larger objects, the concepts that are relevant and the patterns that are found are a combination showing interdependence. The properties of simple things such as particles build up the properties of large things such as congregations of particles, but the very nature of those simple things is itself one which is already constrained by the requirement that it be consistent with overarching concepts that are not themselves about particles. I mean, concepts such as time and energy and entropy. Those three concepts can already be given meaning without reference to microscopic particle physics, and to assert that the microphysics 'just happens' to support the expression of these broader concepts is to make a non-scientific assertion. I mean, non-scientific in the sense of preferring the shoulder-shrug of saying 'it just happened' over the quest to understand exactly what happened and what more can be said about it. In the interplay between the language of particle motions and the language of energy and entropy, neither can claim to be 'more fundamental' or to make the other redundant. It may happen that in the future a more complete particle physics will show that thermodynamics as we currently understand it is not the whole story; indeed, that is what one would expect. But this is not because the interplay of energy and entropy is less fundamental than a theory of the non-composite entities we call 'fundamental' particles. In fact, a future fundamental particle theory will only be correct if it satisfies the demands which are placed on any description of things which can both persist in time and congregate in ways that allow change and development. Those demands are what I am calling symmetry principles (in a natural broadening of the standard use of the term); the first and second laws of thermodynamics, or some more nuanced descendant of them, are examples.

The purpose of this discussion is not to state a straightforward preference for making thermodynamics axiomatic in physics; rather it is to show that one may adopt a more neutral and fluid position, in which both thermodynamics and quantum field theory work together, sometimes one, sometimes the other having the priority, and neither laying claim to having reduced the other to the status of 'afterthought' or 'unnecessary hypothesis'. This has large similarities with, and may in fact be the same as, the position advocated, and expressed in more formal terms, by Butterfield.[15, 16]

In view of this I think that the arguments I have presented so far already go a long way to showing why Figure 3.2 is better than Figure 3.1. The two-way picture of 'support and enarch' both captures the structure of scientific understanding better and also describes the nature of the physical world better. This is true of our best scientific understanding now, and there is no sign of it going away as we learn more. However, it will be important to discuss biology, which I turn to in the next chapter, and also, since the notion that 'physics explains chemistry' is so deeply embedded in the popular psyche (even professional scientists succumb to it), I will finish this chapter with some further pieces of evidence that that notion is sufficiently misleading and inadequate as to be wrong.

Consider the following attempt at a piece of reasoning.

1. An ordinary gas at low density is a collection of molecules whose motions are described by quantum mechanics.
2. These quantum motions result in a gas pressure that falls in proportion as the volume is increased, at any given temperature. (This statement about pressure is called 'Boyle's law'.)

Therefore:

3. This aspect of gaseous behaviour is thus explained in full. No further ideas, such as the laws of thermodynamics, can have anything further to add. Consequently they are superfluous to the goal of understanding Boyle's law.

This attempt at a *syllogism* is a piece of bad logic, and it fails. The failure is of the type called *non sequitur*. The premises are correct, but the mistake comes in at the word 'therefore'. Assertions 1 and 2 are correct, but assertion 3 does not follow from them, and in fact assertion 3 is wrong.

In order to show that assertion 3 is wrong, one can give an argument in which the behaviour of a gas is *derived from* the laws of thermodynamics, plus merely the idea that a gas is made of many small parts. The argument is provided in the Appendix. In such a derivation one does not need to say very much about the motions of the parts that make up a gas. One does not need quantum theory, for example, and nor does one need the detail of any other theory such as Newton's laws of classical physics. One just needs the sort of broader ideas that thermodynamics provides. It follows, and is indeed the case, that the overall behaviour is completely independent of whether the small parts are classical particles or quantum waves. Also, since one does not need to worry about the equation of motion of the parts, Einstein's relativity theory (which modifies the mechanical laws at high velocity) does not matter either. We thus find that, far from having nothing to offer, in this example thermodynamic reasoning can do almost the whole job without any need for any input from what is called 'fundamental' physics!

I included the qualifying 'almost' here because in this argument one does require some sort of notion that a gas is made up of many parts, and the parts are fairly simple and self-contained (see the Appendix for details). However, the point is that one needs very little, almost no, knowledge of what sort of thing each part is, or how it behaves, beyond that the behaviour respects thermodynamic principles. Furthermore, as I already remarked, one may add that the very equations invoked in the reasoning from particle physics were already shaped by the requirement that they respect the symmetries asserted by thermodynamics.

So, which is the more fundamental here, the quantum theory or the thermodynamic statements? The best answer seems to be to refuse to say, but to assert *both* as playing important roles in a valid, in-depth, scientific, insightful understanding of the natural world.

That is what science is like. It is not like the 'ladder of explanation' of Figure 3.1.

Here is another example. It is sometimes asserted that water has the properties it has because it is made of very many small particles called water molecules, and their motions and interactions 'explain' the properties of water. Now, many of the properties of the flow of water are in common with the flow of other liquids at room temperature, such as mercury, bromine, oil, etc. This is 'explained' by the fact that all are congregations of many small particles with mostly short-distance interactions obeying Newton's laws (or, in more detail, quantum physical laws). But the Euler equations that describe fluid flow can be derived from conservation laws (which in turn follow from symmetry principles) without the need for any sort of atomic model. So in what sense is the atomic model explanatory? One might, with equal correctness, say that, after averaging over the small-scale structure, congregations of particles have to obey the same conservation laws as continuous media, and therefore the Euler equations for fluid flow dictate what large congregations of particles must do. The fluid equations therefore explain what the particles do, where the word 'explain' is shorthand for such notions as 'show that it must be the case' and 'bring insight and illumination to the inquiring mind.' Of course the particle motions also bring insight and illumination into what the fluid does. The point is that the explanatory or illuminatory traffic is two-way, not one-way. One can say that a liquid is some smooth flowing 'stuff' that respects the symmetry principles that lead to conservation of energy and momentum, and therefore whatever the 'stuff' is (i.e. the hordes of small molecules), it has to behave accordingly. Equally one may start with the molecules and examine their individual motions and interactions, and confirm that they do indeed have the sort of nature that can give rise to fluid behaviour obeying Euler equations. Both contributions are valid and neither displaces the other. In particular, the Euler equations did not need atomic physics to explain them.

Similar remarks could be made about chemical reaction rates in equilibrium. These are described by an equation called the van 't Hoff equation which forms a standard part of undergraduate chemistry courses, and which is derivable by a thermodynamic argument. A more telling example, perhaps, is that of chemical reaction networks and chemical organization theory. This involves complex networks of reactions in which various feedback loops occur. What is significant is that one can find various overarching patterns of behaviour of such networks that are independent of the details of the specific reactions involved.

By bringing together the equations for pressure in a gas, and the equations for fluid flow, one begins to open up the study of flowing gas, and hence the study of larger things such as weather systems. The latter involves the further complexity involved in condensation of water and heat exchange between the atmosphere and the ocean. The interplay of large-scale and small-scale reasoning continues.

The whole area of non-linear systems, whether in physics, chemistry, engineering, earth science, biology, or other fields, has given rise to many important ideas that describe composite systems and reveal universalities in what might otherwise seem to be disparate systems. This has given rise to scientific disciplines such as chaos theory, the study of critical phenomena, and complexity science.

The following illustration is a less formal statement that serves simply to invite further reflection on the theme of this chapter.

Suppose there were two highly intelligent (but not super-intelligent) beings, A and B, and suppose for the sake of argument that all physical behaviour consists in movements of particles that are correctly described by particle physics. Let A have complete knowledge of the principles of particle physics and suppose A is also furnished with a complete statement of the state of the universe at some time, in complete detail, but with no further guidance concerning what larger concepts make sense. Suppose that B is told simply that the universe is such that somewhere in it, communities of agents can come to be, such that those agents can exhibit rationality and compassion.

Which of these intelligent beings is in possession of a greater degree of understanding, or of a description with a greater degree of explanatory power? I contend that it is not straightforward to say. It seems to me that neither has a very complete understanding, and each has much to learn from the other. However, I am inclined to say that if one had to make do with some partial knowledge, then B's initial knowledge is arguably more insightful than A's. To one who argues that a super-intelligent A could deduce B's knowledge, I reply that, for all we know, it may be possible for a super-intelligent B to deduce A's knowledge, or at least as much of it as is needed for a very thorough grasp of what sort of universe we have. To allow a super-intelligent B to be equally capable of accurate prediction as A it might suffice, for example, to let B also have the microstate specification which we gave to A. (One can imagine that it is possible to explain to B the language in which the microstate is specified without thereby directly furnishing all the details of the equations of motion.)

In all the above examples, the theme of *parts and whole* is repeatedly involved, and an inattentive reader might think I am just rehearsing reasonably straightforward facts about the way large things can result from the combined behaviour of many small things. In fact, as I hope the reader appreciates, I am making a stronger and different assertion.

I am saying that the phrase 'result from' in the previous sentence is often misleading and usually a half-truth.

When one says, of a flow of liquid, 'the liquid flows as it does only because the particle equations dictate that this is what must happen,' one makes a misleading and half-true statement, because one has failed to note that the flow satisfies the symmetries that relate to energy and momentum, and those symmetries tell us much about both the fluid flow and the particle equations. The situation is similar to, or somewhat reminiscent of, the situation in human relations when someone responds illogically to a line of reasoning by invoking the phrase 'you are only saying that because...'. Suppose I make a case that my son ought to go to bed, because he needs his sleep and he will feel better for it, and he replies, 'you are only saying that because you want to watch a film.' The reply is understandable but it is not logical, and it can be at best a half-truth, because, after all, the case I have made (that he needs his sleep) is relevant and correct. An example closer to the theme of the chapter would be if I set out a proof that there are an infinite number of prime numbers, and someone replied, 'you are only saying that because your neurons are excited.' In this chapter we have made a substantial inroad into unpicking the fallacy of that way of speaking. The subsequent chapters will take this further.

4

The Structure of Science, Part 2

The discussion of the previous chapter is extended to the rest of science, especially biology. The *Embodiment Principle* is introduced. This is a statement intended to capture correctly how scientific descriptions interact and interconnect. The concept of *reduction* is discussed, arguing that it does not amount to a replacement of one language by another, but rather a recognition of both languages. I then proceed to biology. Darwinian evolution does not take place in a metaphysical vacuum, but rather explores the space of what is possible. The results of evolution are shaped by various high-level patterns that function analogously to symmetry principles. I give examples. The role of uncontrolled change is such that it enables previously undiscovered such patterns to become embodied.

In the previous chapter I started by setting out the problem, and previewing the conclusion. I then began the argument for this conclusion by discussing symmetry and symmetry-like principles in physics, and making the connection to chemistry. I argued that the notions of *explanation*, *insight*, and *understanding* involve a structure of ideas which includes various languages or types of discourse and is two-way, not one-way. That is, it is not the case that a lower-level language suffices to explain the higher level, nor that a higher-level one suffices to explain the lower level, but that both contribute to a full understanding. Each illuminates the other without making the other redundant.

In this chapter it will be argued that something very similar is going on in all or most areas of science.

4.1 The Embodiment Principle

The present discussion is clearly in the philosophical territory called *reductionism*, but it is not directly addressed to many of the questions commonly posed in that territory. Such questions are concerned with the careful definition, and merits or otherwise, of ontological, epistemological, and methodological reductionism.[47, 21, 26] My main point is that the issue which symmetry illustrates is not clearly brought in view by the questions commonly posed in discussions of reductionism.

Reductionism, in brief, is the idea that a composite entity or a whole can be described in terms of its parts: to specify the whole it suffices to specify the parts and their locations and movements individually. A related concept, called *emergence*, refers to the fact that larger patterns and regularities may be true of an ensemble, owing to the net effect of many parts, without the need for new types of physical interaction (beyond those that the parts show when they are separate or only combined in small groups). Both concepts are widespread and important in science. However, the relationship between an equation of motion and a symmetry principle is not one of either reduction or emergence. Principles such as translational and Lorentz invariance are not the result of the net effect of many small parts.

The increase of entropy (the second law of thermodynamics) might be an example of emergent behaviour,* but this example is so fundamental to the very notion of time that it is a special case. The increase of entropy with time is not a coincidence that could have been otherwise (as one is tempted to say of some emergent phenomena). Rather, it is a way of talking about the notion that entities have what might be called a 'state' or 'identity' which can nevertheless be said to evolve; it is closely connected to the subtle metaphysical questions which arise in discussions of time, identity, past, and future.

In the examples of the car engine, and electromagnetic phenomena in general, we noted that in each case, there is both an equation of motion and a symmetry principle in play. As I have explained, someone might think that the equation of motion somehow undermines or replaces the role of the symmetry principle, but this is quite wrong. In fact, the grander and more subtle idea captured in the symmetry principle is not reduced or modified one iota by the equations of motion. Rather, those equations themselves embody or give physical expression to the symmetry principle.

---

* I say 'might be' rather than 'is' because it is not clear whether or not there is entropy associated with even single-particle processes or with space-time itself.

I think this notion of 'give physical expression to' or 'embody' is a cogent way of describing the situation, and I will be arguing in this chapter that it is a widely occurring theme in science. It is a truth of much of science that has been lost sight of by some exponents of science, and which now needs careful reassertion. To this end, here is a statement that I consider to be helpful:

> **The embodiment principle**. Science is about building up an insightful picture in which the underlying microscopic dynamics do not replace, nor do they explain, the most significant larger principles, but rather they give examples of how those larger principles come to be physically embodied in particular cases. The lower-level and higher-level principles are in a reciprocal relationship of mutual consistency in which each illuminates the other.

I think this may be a helpful way of capturing the sense of 'understanding' that science is always at pains to attain. The assertion which I have called the 'embodiment principle' attempts to express what people practising science often find to be true in their experience and in the considered analysis of how their scientific discipline operates and connects to their wider values. Such a statement is intended to be receptive to the obvious validity of reductionism in large parts of science, and neutral as to whether or not this will be true of all aspects of physical behaviour But it makes a positive pre-emptive strike against the notion that 'reduction' directly equates to either 'explanation' or to 'understanding'.

For the sake of clarity I shall immediately address a thought that might seem to be an objection to the point I just made, but I will argue is not. This possible objection is that there is a well-argued notion employed by philosophers of science, in which the word 'reduction' is used as shorthand for a set of connections between one physical theory and another. The connections are carefully identified (they are called 'bridge principles'[47]) and they show how equations describing the behaviour of some types of composite thing can be extracted from equations describing the detailed behaviour of their parts. Or, such connections show how equations describing a certain type of behaviour can be extracted as a special case from equations describing more general behaviour. These connections are well argued and they certainly earn the adjective 'explanatory', so does this contradict my assertion that reduction does not equate to explanation?

In such discussions, the two descriptions or physical theories that are being connected are like two languages: a higher-level one and a lower-level one. My point is that in order to make these connections, *both* languages have first to be learned. In order to show what is called a 'reduction' in this sense, one must first recognize the nature and categories of the higher-level language, and then note how these are supported on, or are implied by, the terms of the lower-level language applied in the appropriate circumstances. And if it happens that the higher-level language has a certain elegance of its own (as often happens), one may come to consider that the higher-level language has a role in and of itself in an insightful understanding of natural phenomena. It also often happens that the same higher-level language can emerge from several different low-level languages; this again underlines that the higher-level language has significance of its own. If the word 'reduction' is being used in a technical sense as a shorthand for this entire picture, then really the word asserts the validity of *both* languages, not the replacement of either by the other.

The sense of the word 'reduction' that does not constitute explanation or understanding is when one merely asserts 'Oh, the behaviour all follows from L', where L stands for a low-level language, but one has no grasp of the higher-level structures that are in fact present in the behaviour.

In the former case, where 'reduction' refers to the recognition of both levels, then one should note that that which is explanatory is *not located in the lower level alone*. For example, the laws of thermodynamics (suitably generalized to include black holes) are obeyed, as far as we can tell, by systems whose behaviour is described by quantum physics and general relativity (if one accepts one of the strategies to handle the well-known difficulty of proving this). But this does not mean that the thermodynamic laws and their consequences are 'explained' by the equations of quantum physics and general relativity alone, because it is only when one gathers together facts about particle physics in an appropriate way, by taking certain limits and formulating certain useful global concepts, that one can show that thermodynamics is what results on average. *It is this very act of formulating appropriate concepts and taking certain limits* which does large parts of the explanatory work, contributing large parts of the resulting explanation. The original equations that one was working with do not do this. It is work done by a human or other competent agent.

Furthermore, one could always start out from thermodynamics and assess the trustworthiness or otherwise of the particle theory by enquiring whether or not it supports thermodynamics in the appropriate limit, as I already remarked in the previous chapter. At the risk of being repetitive, the next two paragraphs labour this point once again.

For brevity let us use the phrase 'particle physics' to mean the mathematical language offered by a good physical theory such as quantum mechanics, when it is applied to the low-level constituents such as individual particles in a large composite object, and describes those particles and their interactions one by one. Particle physics, so defined, does not explain thermodynamics because it does not even 'see' or 'speak' thermodynamics. The situation is analogous to the example of digital computing presented in section 3.1, and to the examples of symmetry presented in section 3.3.

If a community of scientists knew all the equations of quantum field theory but had never heard of entropy (unlikely as that is), then the eventual derivation of the second law of thermodynamics would amount to a great stride forward in understanding. The idea that microphysics is explanatory but thermodynamics is not can only be maintained by an extremely artificial definition of the word 'explanation'. One can always get truisms by suitably defining words, of course, but one must be careful not to introduce definitions that exclude large parts of what a word ordinarily means, or one will have embarked on meaningless speech. The definition of 'explanation' that would be required in order to support a claim that thermodynamics is not explanatory would be an extremely distorted definition. It would not align with the general concept of *elucidation* and *gaining understanding*, and this is a very severe criticism. Therefore we must conclude either that (a) our speech has to descend into meaninglessness, or (b) thermodynamics is included under the title 'explanation', or more generally, 'what contributes to explanation.' That is to say, it has a role in the explanation of what goes on in the world; a role that cannot be excluded or denied. Neither can the second law of thermodynamics (nor the first, for that matter) be unambiguously assigned a less fundamental role than (for example) Schrödinger's equation.* If one says that the second law is an added item that one does not need once one has already stated the principles of

---

* Schrödinger's equation is the equation that describes the motion over time of basic components of the world, such as fundamental waves and particles.

quantum field theory, etc., then I reply that I may equally assert the second law at the outset and then I will not need quite such a thorough presentation of the field theory, because non-negligible parts of it are implicitly stated already by the axiomatic assertion of the second law. Of course it will be very difficult to know what one may leave out of the field theory in such a way of framing a minimal basic physics (one respecting Ockham's razor), but the point is that in principle it may be done, *and something very like it is actually done in practice when we do science.*

As has already been made clear, the purpose of this view is not to reverse the direction of the supposed 'ladder of explanation,' which would merely be to repeat its mistake. The purpose is to throw out that supposed ladder and realize that the natural world and its scientific elucidation has the two-way structure that I have tried to summarize in Figure 3.2 and illustrate by various examples (including Lorentz contraction, the thermodynamic Maxwell relations, the equation of state of the ideal gas, the Euler equation for fluid flow, the van 't Hoff equation for chemical reaction rates, the theory of critical phenomena, chaos theory, and the organized criticality that occurs in chemical reaction networks). In all these examples the 'explanatory work' is done as much by formulating the right concepts and language as it is by having some good equations of particle physics. And each such cluster of concepts and language lives somewhat loosely in relation to the underlying support, just as the high-level programming language of a digital computer can be supported by a variety of different operating systems and physical hardware.

I have assigned a label (the 'embodiment principle') to the above statement in order to be able to refer to it. A major aim of the present book is to invite people to consider whether or not this assertion captures the nature of scientific explanation fairly and correctly, and may be presented with honour to the next generation—or whether it can and should be rejected. I will also argue that it is important to include a correct view of the structure of scientific explanation as part of general education. We now turn, in the following section, to the discussion of biology.

4.2 Biology

Behaviours and shapes observed in living organisms, social groups, and ecosystems exhibit large amounts of large-scale pattern and structure, and these patterns and structures can be studied in their own right without regard to, or with only limited regard to, microscopic structures and behaviours. It is easy to acknowledge this and it is not controversial. However, there is also a large amount of happenstance in biology, and it is not always easy to tell which features are essentially owing to uncontrolled variation and could easily have been otherwise, and which features could not. This makes it harder to apply the type of symmetry-like reasoning that I have presented in the context of physics. However, I will argue that that sort of approach does in fact give large amounts of insight into biology.

In view of the fact that Darwinian evolution involves an important contribution from uncontrolled variation, it is sometimes asserted that large-scale biological structures are so much the result of happenstance that it is fruitless to try to predict evolution and inappropriate to seek general rules of structure and behaviour amongst living things. It is easy to contradict such assertions, however, because evolution is constrained by what is physically possible. The evolutionary process of life on Earth (or anywhere else) could not produce a bird with negative mass, for example, nor a fish that did not conserve energy. These simple counterexamples are easy to construct, and they already show that it is quite wrong to say that evolution is merely random. But the more interesting question for present purposes is, are many of the more complex patterns observed in biology also unavoidable?

The answer is not immediately obvious because there is such a huge amount of variety in living things. Bacteria, fungi, grasses, ants, baboons, spiders, trees, fish, herons, sharks, earthworms, giant squid, butterflies … the list goes on and on. Some spiders eat their mate after mating is done. Fish generally produce large numbers of eggs, fertilize them, and then abandon them. Whales, on the other hand, exert considerable trouble to nurture their young. Millions of behaviours and forms exist. Underneath this variety there is the remarkable common ingredient that all of these are ways of behaving that can result in transmission of the DNA that codes for them. Seeing this, one might begin to feel that the combination of DNA transmission plus variety (owing to chance variation) is all that can be said. One may, perhaps, add the point I made in the previous paragraph—that the variation takes place within the constraints of basic physical limits—and then stop: 'That's all that can be said; after that it's basically the result of chance amplified and writ large.'

I want to show, in this and the next section, that that view is not guaranteed to be correct; indeed it almost certainly is *not* correct.

I claim that many of the higher-level patterns or structures observed in biology are unavoidable, in the sense that organisms and ecosystems that persist over many generations must exhibit them. An example of such a higher-level pattern is the presence of *sensing and response*, by which I mean that a member of a reproducing species that can persist in a changing environment must have some ability to sense its environment and modify its behaviour in response. I don't know if there is a counterexample to this, but if there were, then I contend that the notion of sensing and response would still be a good insight which applies to the vast majority of organisms.

Now the question arises: in what sense is there a 'must' here? In what sense does the natural world constrain evolution so as to insist on sensing and response? The situation is different to the constraints arising from fundamental physics, such as the restriction to positive mass, conserved energy, increasing entropy, and so on. These prevent various types of structure from ever arising in the first place. The constraint to sensing and response is applied less directly but it is there nonetheless, applied via natural selection. If in fact organisms with a greater adaptability tend to do better, other things being equal, then an ecosystem brought into being by a process of variation and natural selection will be much more populated with that type of organism than with others. The notion of 'sensing and response' is here playing a symmetry-like role, because it asserts what kind of structure and behaviour can make sense in evolutionary biological terms. It is an abstract principle which is true of the search space which the Darwinian process explores.

It is not necessary to the present argument that this example is a valid one, only that there exist valid examples in sufficient number, or playing sufficiently important roles, that the search space of evolution is significantly shaped by them. I think that a strong case can be made for this, though I will not make the case at length here; others will be able to do so with greater expertise than I. I will simply offer a few further examples of things that I think evolution could not fail to discover.

1. The general notion of sensing and response outlined above can be made more specific, because different types of phenomena that can be sensed (e.g. chemical species, pressure waves, electromag-

netic waves) each constrain what kinds of things can sense them.
2. Naturally strong load-bearing structures such as arches and round columns.
3. The preponderance of sexual over asexual reproduction in complex organisms. This is the principle that a reproductive mechanism will proliferate if it allows whatever is the best type and amount of variation from one generation to the next, and in the case of complex organisms this optimum is better approximated by sexual than by asexual reproduction.
4. The principle of 'tit for tat': that is, that organisms do well if they have ability to recognize other individuals and recall their previous behaviour, and act with a preparedness to cooperate, but to withhold cooperation from others deemed untrustworthy.
5. The principle of 'concern for offspring': that is, that when their abilities of recognition and response are sufficiently developed to make it possible, then parents take trouble to enhance the life prospects of their offspring.

The upshot of all this is simply stated. In the end, and unavoidably, Darwinian evolution on Earth could only produce communities of living things capable of living alongside one another on Earth. The capacities required to do so are themselves constrained by all sorts of fascinating high-level patterns and principles (in contrast, the strategy 'behave randomly' is a poor evolutionary strategy). If one does not assert this, then it would seem that one must make the bizarre claim that all these patterns are just sequences of coincidences—hardly a claim that merits to be called scientific. In fact, large parts of the human efforts to understand the natural world are devoted to understanding how biological structures and processes work together in patterned, non-random ways.

Such patterns and principles describe the available search space that is explored by evolution. The principles of ecology, and of social existence, and of sensing, and of a myriad of other areas of biology, all encapsulate truths about what manner of thing an organism is, and they provide concepts in terms of which organisms can be understood. They also reveal to the inquiring mind large amounts of information about what sorts of chemistry and physics could support them. In all these respects they play a role in biology similar to the role played by symmetry in physics.

Suppose we ask a simple question such as 'why are tree trunks of roughly circular cross-section, not elliptical with high eccentricity, nor triangular?'. The answer includes such statements as 'natural selection favours the efficient use of materials', and 'the circular cylinder offers a more efficient use of materials than the highly elliptical cylinder, for a given strength'. The latter is not a statement coming from evolutionary biology per se. It comes from the engineering science of materials and load-bearing structures in general. My point is that this simple example illustrates a very wide-ranging truth about biology; namely that evolution explores patterns that are laid down for it in the very way that things can be, and I assert that this truth can be traced into the rich social and psychological lives of animals, as well as into basic physical facts about them.

It may be objected that my fifth example above (concern for offspring) is not valid, because of counterexamples such as the cuckoo and the shoebill (shoebill storks lay two eggs and then pick one of the two offspring and only nurture that one). My reply is twofold. First, these counterexamples are the exception not the rule in birds; in most bird species parents invest a great deal in their offspring. Second, in cases where such nurture does not happen, the degree of recognition of connection is a lesser one. The pattern here is the combination of *recognition of connection* with *effort to nurture*.

Taking this same point further, one may observe that in complex animals one encounters either loners who fend for themselves, or else one encounters social groups. But if there is a social group (a pride of lions, a troop of monkeys, etc.) then it must realise those behaviours which are necessary for a social group to function. It is tautological. By definition, if the group does stick together and thus get better prospects for the next generation, then the group members have to live in such a way that this is what happens, and this entails that to live in complete ignorance of other members, or in complete suspicion of them, cannot be the order of the day. A bit of recognition and trust is required. This is not mere randomness writ large; it is the expression of interesting and deep connections; truths that are well worth seeing and pondering.

4.2.1 *The Evolutionary Development of the Brain*

The human brain is an example of this general principle.

The human brain is a product of Darwinian evolution, but this does not imply that its capacities are arbitrary, nor that they have very much to do with survival and reproduction. A good example of this lies in our ability to follow mathematical logic. If various mutations had gone otherwise, would it be the case that we humans would consider that one plus one equals three, or that the area of a rectangle is other than the product of the length of two of its sides? Of course not. Either we would not understand integers and arithmetic at all, or we would agree that one plus one makes two. Basic mathematics is not a product of evolution; it is a set of logical principles and truths about sets. It is one among the abstract languages that thinking physical beings such as ourselves can come to appreciate.

The abstract logical structures that we call mathematics represent a language that satisfies its own internal rules. This language and its rules are not products of anything physical, nor the result of any purely physical process. Any thinking beings that come to exist in the universe will either grasp something of this language or they will not. But if they do begin to grasp logic and mathematics, then that is what they will grasp. This is a tautology, but an important one. We have here another mould or constraint, and one that acts at a highly subtle level, shaping what the possibilities are for the products of a physical process, whether Darwinian evolution or any other. If a gradual process, or any other type of process, results in thinking agents capable of abstract logical thought about number, then that thought will perforce be logical and about number. Such agents will, for example, consider that five is a prime number and six is not. Evolution is utterly constrained by tautologies of this kind. It can only result in agents that either do not understand maths, or do understand maths, but if they understand maths, then they understand that six has factors other than itself and one, and five does not.

Our very identity is profoundly shaped by this sort of fact. We are not the products of low-level physical evolution merely; we are the products of high-level truth, of what *is* or *can be*.

4.3 The Role of Uncontrolled Change

One of the reasons that the 'ladder' picture of science, and hence of the natural world, is so hard to dislodge is that people have a sneaking suspicion that the world is like a vast machine, such that in fact, no matter what we discover about biology, climate science, neuroscience, and so on, we will just be filling out some observations about what the machine throws up by the operation of its basic working principle, and all the further observations are merely statements about what that basic working principle happens to do. On this view the further patterns have no traction on truth and reality in and of themselves; they are wholly subordinate to various impersonal low-level laws that are called 'fundamental'.

This is wrong, for the following reason.

The processes of the natural world are either deterministic or they are not. That is, either there is no element of uncontrolled, random change, or there is such an element. There is plenty of evidence to suggest that the situation is the latter one—and that is what I think—but let's look at both possibilities.

If the universe is deterministic then the presence of all the structure it now has can be traced back to its initial conditions. On this view, the structure now present at every scale has a direct correspondence with the presence of structure in the early state. It is not the working principle alone of such a machine which gives the universe its present form, but rather that working principle getting to work on the very special initial state. If we compare this to the working of a deterministic computer, then it is as though there were a very special program stored at the beginning. The output that is produced later on is very much a product of that stored program.

Now let's consider the other possibility: a certain amount of non-determinism or openness.

If the universe is not completely deterministic, then there is not such a direct link between its present state and its early state. The presence of uncontrolled change reduces that link. The evidence from physics is that there is such uncontrolled change, and it takes the form of a continuous low-level random fluctuation. In this case, the evolution of the universe can explore a range of possible eventualities, and what will happen is that *those sorts of relations and concepts that can in principle make sense may have a chance to come to be physically expressed.*

Universal uncontrolled low-level fluctuation acts like the shaking of a sieve, or, a better analogy, it may be compared to the way that shaking a complicated mould will cause powder or paint to more completely run into and fill the patterns and corners of the mould. The 'mould' in this case, however, is existence itself, the very meaning of the verb *to be*.

The shaking enables things in the universe to find out and to fill or express what can *be*. Here are some examples.

1. Define a sphere in three-dimensional Euclidean space as a surface formed by points equidistant from one other point (the sphere's centre). All sorts of nearly spherical objects can come to exist, but none of them will fail to have rotational symmetry to good approximation. This example concerns the physical embodiment of a mathematical truth.
2. An efficient concave dam wall can be, but an efficient convex dam wall cannot be, when working with materials of higher compressive than tensile strength. This example concerns a truth that is partly mathematical, but also involves some general physical concepts such as stress and pressure.
3. A sense organ for light (i.e. visible electromagnetic radiation) must be an organ that can react in some non-negligible way when light hits it. This example concerns the relation between one sort of high-level structure and the nature of a basic physical phenomenon.
4. A social animal must be a creature whose cognitive abilities confer some sense of what social interactions are. This is a statement about the meaning of the word 'social,' and how that meaning can be physically expressed and embodied.
5. A fair division of labour without slavery can be, but a fair division of labour with slavery cannot be. This illustrates how a basic truth about fairness can be illustrated and embodied in arrangements for human life.
6. A rich person who finds it hard to enter into life at its most meaningful and humane can be, but a rich person who finds it easy to do this cannot be. This is because the very notion of ownership and material riches is itself distorting, when it comes to relationships amongst persons.

The last example is a statement that not everyone will recognize as true, or whose truth is hard to establish; I include it merely to illustrate that subtle considerations of this kind are included among the set of things that are or can be.

The fact that the above examples, and an infinite number of further ones, can be physically realised is owing in large part to the innately non-self-contradictory, almost tautological, nature of each such example. The role of the underlying working principle of the physical world is not to dictate all these outcomes, nor to announce that they are all meaningless, but to provide that combination of reliability and flexibility which makes it possible for a rich variety of such truths to come to be embodied.

If in fact the ongoing evolution of the universe is not completely controlled, and the non-deterministic model is correct (as the evidence suggests and as I strongly suspect), then that very lack of complete control is a strong contributing factor to the ability of the physical world to embody and express rich truths of many kinds. It has enabled Darwinian evolution to have the freedom it needs in order to discover and invent. It is in part owing to this freedom that there can be cells, and insects, and mammals such as ourselves.

The freedom under discussion here is often described as 'randomness', and the presence of randomness is often regarded as a defect in the grand scheme of things. But I would say that it is not, even though this same randomness also opens the way to suffering, through cancerous diseases, unpredictable weather and violent avalanches, old age, and so on. My aim here is not to assess and pass judgement on the goodness or otherwise of the physical world, however (a role for which I am ill-equipped). My aim is to comment on the fact that a certain degree of randomness, or openness (these are two words for the same thing), is a feature that allows high-level languages to come to be discovered and expressed in the ongoing evolution of the physical world. That world is not a mere mechanism going through meaningless motions; it is a world which develops in its ability to encounter and express deeper and richer forms of truth.

5
Logic and Knowledge: The Babel Fallacy

I describe a certain error of logic, which I dub the *Babel Fallacy*. This is the fallacy of claiming that one already knows that a given low-level language is adequate to support the expression of a given high-level or collective phenomenon, when this has not been shown. Examples from physics, linguistics, economics, mathematics, and computer science are given. The same reasoning applies to biology. In the context of science, the Babel fallacy is the fallacy of claiming to know the complete truth about the physical nature of anything.

I now wish to bring out a further aspect of the sense in which biology gives insight into physics, and to draw attention to a significant logical fallacy.

Consider the behaviour of an ordinary mammal such as a monkey. I will not try to describe the behaviour in general, but rather use one feature of it in order to make a simple point that is largely about logic.

Suppose that someone says that a monkey is a collection of cells, and that cells are collections of molecules. So far, so good—we can agree that; it is true. But suppose someone adds some further phrase, such as, 'and that is all': 'It is a collection of molecules, and that is all.' The logic is to do with the meaning of the phrase 'and that is all'. Such an added phrase might be true or false, depending on what is meant by it. If it is serving as a way to exclude the presence of physical substances unknown to physics, then it is a reasonable thing to say. However, if it is serving as a way of saying that low-level explanations do all the explanatory work in science then it asserts an untruth, as I have argued in previous chapters. In this chapter I will bring out a further sense in which a phrase such as 'and that is all' can be untrue because functioning as a veiled way to assert an untruth. The point is readily made by analogy with a simpler physical system.

Logic and Knowledge: The Babel Fallacy

Consider helium gas, for example. Someone may say 'it is a collection of atoms', and they would be right. But suppose we cool the helium gas down. 'I know,' they will say, 'it will liquify: it is a collection of atoms.' Well yes, but suppose we cool the liquid down some more. 'It will solidify,' our friend will say. Perhaps they may add, 'It is well known that liquids turn to solids when you cool them, and this is because they are, in the end, collections of atoms, and that is all.'

Yes, but in fact helium does not solidify at one atmosphere of pressure when it is cooled down. At low temperature it becomes practically the opposite of solid: it becomes superfluid. And that is a surprise. Yes, it is a collection of atoms, but this collection becomes superfluid below a certain temperature (called the *lambda point*), and this was totally unexpected.

The logic here is the following. When we say that helium is a collection of atoms, we are right, but the danger is in assuming that we already know all about what atoms are and what collections of them can do. Because then we would be wrong. We don't know all about what collections of atoms can do. We have very limited knowledge of what they may be capable of. Of course, by now we have got a pretty good understanding of superfluidity in liquid helium, but the collection of atoms that is shown in Figure 5.1—the monkey—that is much more complicated. And here is the point:

Claim. *When we say that a monkey is a collection of atoms and molecules, we are right, but that does not logically imply that we understand monkey behaviour, and furthermore, neither does it imply that our current best understanding of molecules is sufficient to support*

Figure 5.1. A monkey.
Source: Billion Photos; shutterstock 293857409.

a correct model of monkey behaviour. In fact, it is the monkey behaviour itself that will teach us what monkey-shaped collections of molecules can do. *In this way, the monkey fills out our understanding of what molecules are, at the same time as it fills out our understanding of what monkeys are.*

Now I want to underline the strength of this assertion. The point is, we don't know (fully) what anything is, whether monkeys or molecules. We need to combine all sorts of observations in order to learn what such things are. Studies of molecular physics and chemistry give plenty of support for an algorithmic model of molecular behaviour, for example (it is the model known as quantum mechanics).* It is widely assumed that that algorithmic model, when applied to something as complex as a monkey, could give us warrant to say that the monkey-shaped collection of molecules has genuine concern for its offspring, and not just the appearance of this. But what if such a model could not do that? What if there were a convincing argument to show that an algorithmic model could not give rise to things like genuine concern, but only a sort of pretence? What would be the logical conclusion? We have two options: to deny that monkeys can express concern, or to deny that we had understood molecules in full. My point is that the former option, which gets widely touted as 'scientific,' is in fact the unreasonable and, indeed, unscientific option. The reasonable option is to allow the experimental evidence (superfluidity in liquid helium; social behaviour in monkeys) to alert us to the defects in our understanding of the physical apparatus (atoms and molecules).

To repeat, *it is the observed behaviour that tells us what molecules can do* and therefore *it is the observed behaviour that tells us whether our understanding of molecules is in fact correct.*

In the context of scientific explanation, the phrase 'and that is all' is untrue whenever it is employed as a way of implying that the speaker or listener already knows what higher-level patterns or languages can be supported on or expressed by the lower-level apparatus under discussion, in situations where this is not the case. This is a different untruth to the one I discussed in previous chapters. They can be displayed separately as follows.

---

* By an *algorithmic* model I mean the sort of mathematical recipe that one could imagine programming a digital computer to calculate (all models currently employed in chemistry are like this).

Logic and Knowledge: The Babel Fallacy

The following assertions I hold to be true:

1. Low-level explanations do not do all the explanatory work in science.
2. In the history of scientific endeavours by the human race, the attempt to predict collective phenomena from equations describing the parts has often failed.
3. One is not in a position to know that a given low-level model can support a given high-level collective phenomenon until this is explicitly demonstrated.

In previous chapters I discussed the denial of the first item in this list; in the present chapter I am discussing the denial of the third item. I am also pointing out that such a denial is often not made explicitly, but is implied by an injudicious use of reductionist language.

This type of mistake is sufficiently common, and sufficiently important, that I think it useful to give it a name. Let us call it the *Babel fallacy*. This name is a reference to the myth of the Tower of Babel, described in chapter 11 of the book of *Genesis* in the Bible. That myth is widely misconstrued as a type of just-so story about the origin of languages in human history, but that reading completely misses what the story is really doing. One gets a powerful insight into the story of the Tower of Babel by noticing its similarities with and differences from other stories circulating in the Near East at the time when it was written. At the time, both the Babylonians and the Sumerians held their own version of a story involving a tower and their conception of a god or goddess. In each case the tower is raised in the ruling city, and the divine being takes up residence at the top. These stories were attempts to assert one kingdom or another as the centre of religious and political power. In the Sumerian version, there is a reference to language: the result of raising the tower will be that the 'many-tongued lands' will 'address Enlil [the name of the chief deity in Sumerian religion] together in a single language'.[75] In other words, 'we the Sumerians are Top Nation and everyone else will look to us and speak as we do.' In view of this, one may deduce that the Hebrew version is a deliberate attempt to subvert the existing stories. The Hebrew account of the Tower of Babel is a polemic written to oppose the use of religion to assert the dominance of a ruling tribe and a ruling language. The substance of it is, 'far from capturing God, your tower will be thrown down by God, and far from us all having to speak your language, both we and you are going to have to speak a hundred different languages.' It is the point about language that interests me here.

In the present discussion, applied to natural science, what I am naming 'the Babel fallacy' is the fallacy of supposing that one already knows that a given low-level language can support the expression of a set of high-level or collective physical phenomena, when this has not been shown to be the case. Thus it is a fallacy about overestimating one's state of knowledge.

The above examples (liquid helium and monkey behaviour) already illustrate the point. I will give further examples, but before doing that, I would like first to emphasize that the logical term, 'fallacy', is correct here, because this is a failure of logic and of reason. A fallacy is involved when someone draws a conclusion from a set of premises, but in fact one or more of the premises is false, or the conclusion does not follow from the premises. This is not the same thing as saying the conclusion is false in itself. The concluding statement in a fallacious argument may or may not be true; if the argument is fallacious then the argument simply does not help us to determine the truth or otherwise of the conclusion. To see this (it is familiar to logicians of course), consider the following two arguments, both of which are fallacious:

Argument A
Major premise: For any numbers X and Y, if the even multiples of X are also multiples of Y, then all the multiples of X are also multiples of Y.
Minor premise: Every even multiple of two is a multiple of four.
Conclusion: All even numbers are multiples of four.

Argument B
Major premise: Same as for argument A.
Minor premise: Every even multiple of four is a multiple of two.
Conclusion: All multiples of four are even.

In this example, the conclusion of argument A is false, and the conclusion of argument B is true. However, this does not alter the fact that in both cases the argument is worthless, as an argument, because the major premise is false. This illustrates the fact that when we say that an argument involves a fallacy, we are making a comment on the structure of the process of reasoning, not on whether or not the conclusion is true.

As I already said, the Babel fallacy is the claim that one already knows that a certain low-level language is adequate to support the expression of a given high-level or collective phenomenon, when this has not been shown to be the case. In the example of liquid helium, the fallacy was not to suppose that helium is made of component parts called atoms; the fallacy was to suppose that one already knew what these atoms were like in full. The fallacy could be, for example, to suppose that atoms always behave like individual distinguishable entities. In fact they do not, and superfluidity is related to this fact. In the example of the monkey, the fallacy is to suppose that we know that modern molecular physics is adequate to support the expression of whatever high-level languages correctly describe monkey behaviour.

In the history of science, there have been many examples of the Babel fallacy. One of the most famous is the proposal made by various commentators at the end of the nineteenth century that physics was close to finished; Newtonian mechanics plus classical fields would suffice to describe everything. How wrong they were! This example included two major parts: the incorrectness of Newtonian gravitational theory, and the incorrectness of Newtonian mechanics more generally. Both examples are useful because they illustrate that when a low-level language is inadequate, it can be profoundly inadequate. The move from the Newtonian model of gravity to the one provided by Einstein's theory of general relativity is not just a tweak or a few added details; it is a major shift of the very terms of discourse. The same can be said of the move from classical to quantum mechanics.

The Babel fallacy can also be illustrated by a more everyday example. Anyone who played with Lego building-blocks as a child, or as an adult for that matter, will have appreciated their versatility. You can make a lot with Lego: buildings, aeroplanes, animals, cars. One can make moving parts of great sophistication. In view of this, one might begin to feel that one could make practically any machine. The 'low-level model' of Lego can, one might suppose, support the expression of larger things such as powerful rockets, aeroplanes, and racing cars. In fact, however, when one tries to build an internal combustion engine, or a jet or a rocket, one finds that it does not work because the plastic Lego bricks start to melt (one also encounters other issues to do with tensile strength). The high temperatures cannot be avoided, in an engine powered by burning fuel, if the efficiency is to be sufficient to make a practical vehicle. This situation illustrates the Babel fallacy: ordinary Lego offers a versatile set of parts and connections, but one would be wrong to suppose that it can be used to make machines that require high operating temperatures.

The Babel fallacy can occur in a more formal setting in linguistics. The *Chomsky hierarchy* (also called the *Chomsky–Schützenberger hierarchy*) is an important basic idea in linguistics and in the formal languages of computer science. It consists of a set of four types of grammatical structure, each of which can give rise to sequences of symbols called languages. The types of grammar become more sophisticated as one moves up the hierarchy, and it is a hierarchy because each class of grammar in the list contains the one before it. This bears on fundamental physics as well, because different types of physical theory can be shown to be associated with different types of grammar in the hierarchy. In this context, the Babel fallacy consists in claiming that a given language can be produced by a given low-level grammar, without providing an explicit demonstration.

Mistakes that are similar to the Babel fallacy, though less formally provable as such, are commonly made in economics and political theory. For example, economic theory has often been predicated on fairly simple models of the behaviour of economic agents, such as the notion that agents always act so as to make what monetary gains for themselves they can, under the constraints of what information they have. Such models are useful, up to a point, but they have regularly failed, sometimes with disastrous consequences. In fact economic agents such as people often have other concerns, such as fairness and a sense of identity, which make them behave in other ways. A related example is the overly simplistic treatment of history, such as the one proposed by Karl Marx.

Something like the Babel fallacy also contributes to other human ills, such as racial prejudice. A really severe racial prejudice includes, as a basic component, a simplistic model of human relations in which one race of humans is held to be innately superior to another, and evidence to the contrary is suppressed or ignored. I detect the Babel fallacy here because it is the very language adopted by prejudice, and the thought-space associated with it, which is incapable of supporting the expression of the full richness of human relations.

For my final examples I will return to mathematics and computer science.

At the turn of the twentieth century, it was widely believed by mathematicians that mathematics itself was drawing to a certain sort of completion, in the sense that it would soon be possible to find a way to prove the logical consistency of the axioms of arithmetic, and to show that a wide range of mathematical problems could be solved by an automated (i.e. algorithmic) method. In this way one could automate theorem-proving in much of mathematics, so that all the lengthy proofs that mathematicians embark on could be carried out by machines. In a famous lecture in 1900, the mathematician David Hilbert put these ideas forward, along with twenty or so others, as something the community should try to prove.

Hilbert's second problem asks for a proof that the axioms of arithmetic are logically consistent. In this context, 'logically consistent' means that no sequence of allowed operations of arithmetic can set out from a truism, such as '$1 = 1$', and deduce a self-contradictory conclusion, such as '$1 = 0$'. It seems, to an informal intuition, that it is highly likely that no such contradiction will arise, but in mathematical logic one is not satisfied with informal intuitions. One wants a proof that one's system of ideas is not going to lead to nonsense. It seems amazing that this was unknown, because, after all, most of mathematics relies on the ideas of arithmetic one way or another, so not knowing whether or not arithmetic can be trusted amounts to not knowing whether or not almost any mathematical result can be trusted. This is because if one can derive a contradiction then one can contradict practically anything; the whole edifice begins to fall apart. It was natural to assume that arithmetic does not contain some hidden inconsistency, and most mathematicians believed it would be possible to prove that indeed it does not. Such a belief is an example of the Babel fallacy. The fallacy here is not in thinking that arithmetic is consistent (one may simply trust that it is consistent and there is nothing wrong with such trust); the fallacy consists in believing that the rules of arithmetic can themselves be used to prove that they will not lead to contradictions. In a famous paper of 1931, Kurt Gödel showed that no such proof is possible.

In order to try again to get a proof of the logical consistency of arithmetic, one can bring in further ideas, and if one can simply trust the further ideas then Gödel's result does not rule out that the proof might be possible. But now one has the issue of whether or not the further ideas are guaranteed to be correct. This leads to an infinite regress, and at the time of writing there is no consensus, amongst the mathematical logic community, on whether or not the rules of arithmetic have been, or even can be, proven to be trustworthy. There is plenty of evidence that they are trustworthy, but this is evidence of a different kind to mathematical proof: they are trustworthy because if one trusts them then one experiences the delights of mathematical beauty and the benefits of science.

Another of Hilbert's problems, the tenth, concerns further ideas related to mathematical proof, and the Babel fallacy arose again. It was believed by many that a certain reasonably basic kind of mathematical problem could in principle be automated, so that one could program a computer to solve it, but it turns out that this is impossible. Once again, Gödel's work is relevant. One learns from this to be more cautious about claiming what may or may not be achievable within any given area of discourse.

This leads naturally to my final example, which is the shift in computer science from classical to quantum computing. I already alluded to the paradigm shift from classical to quantum in fundamental physics. The shift in computer science is related to this, but merits its own consideration.

The foundations of computer science go back to work by Alan Turing, John von Neumann, Claude Shannon, and others, who in the mid-twentieth century formalized the notion of information and information processing. The *Turing machine* is a type of machine which captures, in a simple way, the essential features of what computers do. It can be shown that a Turing machine can, in principle, perform all the tasks commonly associated with computers, and this is such a rich range of tasks that the computer science community began to feel that, through the notions of binary logic and the associated logical operations, they had captured, in full, what information processing is. Then quantum computing came along.

This is another example of the Babel fallacy, in the following sense. Quantum computing is a form of information processing which makes use of the full range of behaviours of physical entities such as photons and atoms, insofar as our current grasp of basic physics enables us to learn what those behaviours are. In particular it makes use of those features which require for their correct description the theory of quantum physics. I will not set out the technical details here, but summarize the salient points.

The move from Turing machines, now called 'classical' computing, to quantum computing does not result in a complete transformation of computing, because, in principle, a classical computer (what one might call an 'ordinary' computer*) can be programmed to simulate or mimic

---

* Roughly speaking, a classical computer is one based on digital logic and operations between entities with localized properties, or whose operation is formally equivalent to this.

a quantum computer. However, as far as we know, in order to perform such a simulation, the classical computer will need vastly more physical resources (e.g. more time or more energy or more memory, or all of the above). I say 'as far as we know' because it is not known how to prove this, but the evidence strongly suggests it, and in any case it is certain that the way we understand quantum computers involves a different basic set of ideas to the one used in classical computer science. The sets of ideas are not utterly unlike, but they are changed in profound ways. It follows that the strong intuition, formed by many computer scientists before the 1990s, that classical information processing is the whole story about information processing, is an example of the Babel fallacy.

This concludes my set of examples.

Let us return now to the monkey with which we started. I will state my own (subjective) opinion in the case of the monkey and the atomic model. I do this in the interests of clarity. Note, this is not a case of knowing the answer. It is case where, in view of the Babel fallacy, one should not claim knowledge which one does not have. But we can share our hunches, and here is mine.

I would say that I am as close to certain as makes no difference that a monkey is made of atoms that are correctly described by some sort of fundamental physical theory, but I am also close to certain that the current Standard Model of particle physics is not the whole truth about particles and fields, and I am not at all sure that we understand fundamental physics well enough to allow a good working model of anything like a brain. The whole truth about the natural world is certainly different from what we currently know, and it is probably different in ways of great subtlety, and an area where such subtleties might be relevant is that of physical effects underlying brain function.

6

Reflection

> A reflection on the argument of the book so far, and its application to some aspects of human life. It can, and should, be deduced that the moral stature of human beings is neither eroded nor replaced by the study of the physical mechanism of human beings. It is the duty of scientists writing for the general public to make this clear and not fudge it. It is false to say that the language of justice and responsibility is less objective than statements about wavefunctions.

I will now repeat the main conclusions of the argument of the preceding chapters, and then briefly extend the discussion to some aspects of human life.

One main conclusion is the *embodiment principle* given in Section 4.1. This is a statement both about the structure of scientific explanation and about what is going on in the physical world. It complements Figure 3.2. The figure introduced the general idea of a network of interrelating areas of knowledge and of physical expression; the embodiment principle states what the relationships in this network are like. I presented the application of this principle to physics and chemistry in Chapter 3 and I explained the application to biology in Chapter 4. I have not devoted separate sections to further areas of scientific discourse, such as engineering, materials science, zoology, neuroscience, climate science, psychology, and so on, because I have used the three standard names to stand in for science in general, and I think their discussion suffices to capture the themes I have addressed. Chapter 5 then described the Babel fallacy, which is an error of reasoning that can arise in any area of science.

Among other behaviours, monkeys take some trouble to enhance their offspring's life prospects. According to the arguments I have given, such behaviour is *neither understood in full nor explained in full* by observing that it is an outworking of various mechanisms based on genetics, fitness, natural selection, chemistry, geophysics, oxygen metabolism, and so on. The reductive observation is useful, and has its place in a full understanding, but it does not on its own constitute an *understanding* of the behaviour, because it does not grasp what the behaviour amounts to, what it enacts. The language of *concern* and *agency* is needed to grasp (i.e. formulate an understanding of) that. The explanation of the workings of a monkey offered by a purely molecular or cellular description is such a blind and dumb (i.e. unable to see or speak) sort of explanation as to amount to almost no explanation at all.

Furthermore, it is not true to say that the low-level description is the driver of the whole description, nor is it the more fundamental part. What we have in a biological example such as a monkey is a large set of interesting and insightful biological principles that operate in a similar way to symmetry principles. They include things such as care for offspring, and hunger for food, and occasional territorial aggression. The monkey is the physical embodiment of these principles, in somewhat the same way that a stick is a physical embodiment of translational symmetry and Lorentz covariance (and other symmetry principles of fundamental physics). A collection of creatures able to form a social group could no more avoid signalling and attentive behaviours than they could avoid having positive gravitational mass and slower-than-light motion. When we notice that a monkey is made of molecules, and molecules interact via fields that obey Maxwell's equations, we are not discovering the truth or otherwise of the larger principles, nor anything but very small amounts about how they function. Rather, the larger principles simply are what they are, making their own sense already. By discovering molecular structure we do not nullify or unpick or negate or make redundant the truth of the concern for offspring that is expressed in monkeys, just as electromagnetism does not nullify or unpick or negate or make redundant Lorentz covariance. Rather, the truth of concern for offspring is already there, an abstract principle that makes sense already, without any mention of molecules. It is a truth that shapes what sort of thing a monkey can be, in somewhat the same way that Lorentz covariance shapes what sort of thing the electromagnetic field can be. It shapes what kind of monkeyness can make sense.

This point need not, and does not, deny that the social behaviour of monkeys is somewhat shaped by the particular biological hardware that supports it. But to a large extent that social behaviour follows rules and patterns that are independent of the nature of its support apparatus, just as the equation of state of an ideal gas is independent of much of the underlying microphysics. In short, the social behaviour of monkeys makes sense in terms of its own appropriate high-level language, largely irrespective of what the cellular mechanism is. The behaviour was arrived at by evolution because it is a behaviour that can work—it is one that can enable groups of animals with similar genotype to maintain a presence in the world in the context of other groups of animals doing likewise.

One cannot always argue directly from concepts appropriate to physics to ones appropriate to biology. Consequently, I have presented arguments on why it is valid to draw an analogy between structures relevant to biology and symmetry principles in fundamental physics. My purpose is to draw attention to a similarity, not to an identity. But this similarity is sufficient to make the following point: science simply is not what many widely received presentations of science have claimed. When some exponents suggest that animal behaviour can be described as essentially about genes, or about neurological chemistry, for example, then they are being illogical. They are committing the type of non-sequitur that I illustrated in the failed syllogism presented in section 3.4 (Boyle's law for a gas). Such exponents will, I guess, reply that they do not need some such failed syllogism, because they are arguing from a more careful sequence of reductive steps. My point is that the nature of that very sequence of reductive steps, and what it does and does not achieve, is itself open to a more considered reading, and the very terminology adopted in words like 'reduce' itself biases the mind in ways that are not always insightful. A statement such as 'this set of molecules moved that set of molecules from A to B owing to the chemical reactions in this third set of molecules' does not capture the whole truth, nor does it amount to an explanation. The whole truth involves that one also understands and asserts, 'this set of molecules is the hand of the creature called monkey, that set of molecules is food to the monkey, and the third set is the monkey's brain, which is the physical computer of monkey computations' (a statement intended to be neutral regarding all mind–body problems and the status of consciousness in monkeys).

I am, for the purposes of the present discussion, maintaining neutrality over whether or not reductionism always applies* (reduction-

---

* But it is worth noting that reductionism certainly is questionable when it comes to fundamental physics and quantum entanglement.

ism being, for present purposes, the notion that a description of a whole can be furnished by describing the parts and their locations and movements individually). Even supposing that each and every development in the physical world was in one-to-one correspondence with a movement of fundamental fields or particles, and that one was willing to accept fundamental physics as a given, low-level descriptions would still fail to constitute full explanations, because they would fail to give full understanding. As soon as the language changes level, it has already abandoned some insight. Pointing up this sort of mistake is difficult, because it is not a case of a direct contradiction, but of a lack of awareness that the whole structure of scientific explanation has not been appreciated correctly. A correct grasp of science involves such an awareness.

It would still be false to say that an arch is explained by what stones are, even if all arches everywhere were made of stones.

So, according to the meaning of the word 'explain' that, I contend, is close to its meaning in ordinary discourse, it is false to say that the natural world can be explained by fundamental physics, in the same sense that it is false to say that an arch is explained by the properties of stones. It would still be false to say that an arch is explained by what stones are, even if all arches everywhere were made of stones. This is because it is not essential to what an arch is, and why it has good compressive strength, that it be made of stone. In a similar way, even if the current Standard Model of particle physics were the so-called theory of everything that physicists (quite rightly and honourably) try to discover, it would not itself express what even a humble arch is, let alone a cheeky monkey. To be alert to this requires that one holds an awareness that is demanding on the whole range of our faculties of understanding, but that awareness is what constitutes understanding. It is unavoidable (as well as interesting).

The argument I have offered has been advanced in places by example and invitation, rather than by a reasoning process that follows with the same inexorable logic that a mathematical proof would have. I have invoked translational invariance, and Lorentz invariance, and reflected on what sort of role they play, and I have invited the reader to think in similar ways about the more complex but nonetheless recognizable high-level principles that are expressed in the behaviour of animals. This is not a false analogy, I contend, because there is a sufficient similarity between the two cases. However, I admit that this discussion does not offer proof so much as an invitation. I am showing that it is possible, and it is reasonable, to think of rich physical behaviour in this way. It is a way in which we can respect the genuine insight and sheer truth on show at each layer of meaning that we discover in the world, and in which the lower layers do not unpick that meaning nor overturn it nor replace it. It is a humane philosophy of science as well as one having strong credentials.

So I am inviting the reader to see science a certain way. This is a perspective that has not here been deduced by a process of logical proof, and I suspect it is not amenable to logical proof because it functions at the level of being axiomatic. It is amenable, however, to a sort of empirical proof, in that it promotes fruitful ways of thinking about the world. What I claim is that this is an intensely truthful, reasonable, and insightful way of seeing what science is, and of seeing what the world is.

If we get this right, then we restore to each aspect of the physical world a proper recognition of its nature. We restore to the electromagnetic field its symmetry, and to liquid helium its collective wavefunction, and to monkeys their monkeyness, and to humans their humanity.

In the twentieth century, science proceeded by leaps and bounds, and it shows no sign of slowing. This is a tremendously good thing, but also a dangerous thing. It is dangerous because it puts more and more power into the hands of small groups of humans, and humans are not altogether wise, peaceful, and trustworthy creatures. Some of this impacts on politics, some on global warming, some on arms manufacture and unfair trade, and so on. I will not try to comment on all of these. But I do want to draw attention to another danger, one which is close to the theme of this book. This danger is the danger of using science as a way of promoting a misguided view of complex physical things. It ultimately becomes the danger of objectifying people, and thus dehumanizing people. Here, by 'dehumanizing' I mean the failure to see and acknowledge people in their full and rich reality as human beings, as the humans that they in fact are. Such a failure is one that reaches out of merely academic discourse into the life of the community as a whole, and this makes it very significant. Among other things, it causes deep-seated psychological suffering in some of those who apply it to themselves, and it feeds into the process by which people get drawn into systems of extreme violence.[30, 53, 59]

Consider, for example, the following attempt at a piece of reasoning.

1. Human bodies are collections of cells which in turn are collections of molecules.
2. These molecules and cells behave in accordance with various laws of physics, chemistry, and biochemistry, and their particular arrangement is a product of Darwinian evolution.

Therefore:

3. The notion of moral good and bad is irrelevant to the behaviour of human bodies. The behaviour is caused by the underlying physical and chemical processes, as described by quantum theory and particle physics, and that is all there is to it.

Some readers will be thinking at this juncture about the evolution of co-operative behaviour, along the lines described by Robert Axelrod, for example, in his celebrated discussion of the iterated 'prisoner's dilemma'.* Studies such as this show how standard evolutionary biology can lead to feedback loops which amplify cooperative behaviour. After many generations, this can lead to animal populations in which most individuals are inclined to cooperate with others whom they can recognize as like themselves, and all this happens without any consideration of the morality or otherwise of any action or motive. A philosophically naïve reaction is to conclude that morality is irrelevant. However, such a reaction is illogical: it is, once again, the same sort of logic (or lack of logic) that declares that buildings can be understood without mentioning arches. Just like the example concerning Boyle's law that we considered in Section 3.4, the above attempt at a syllogism is simply bad logic, and it fails. It is a non sequitur. The premises 1 and 2 are correct, but assertion 3 does not follow from them, and in fact assertion 3 is both dubious and possibly completely incoherent.

In the above example, concerning morality, there is a giant leap involved, that the astute reader will be aware of and rightly cautious of. There is leap from one area (scientific modelling and the structure of physical processes) to another (moral judgement, moral good and

---

* The *prisoner's dilemma* is a scenario considered in game theory, in which each of two players has to choose an action without knowing what the other will choose. The payoffs are such as to create situations in which cooperation is mutually advantageous, but if one cooperates and the other refuses, then the latter gets the immediate advantage. In his paper, and book, *The Evolution of Co-Operation*,[11] Axelrod showed that if the game is repeated many times and players can take into account what other players have done previously, then a small amount of initial willingness to cooperate can open the way to widespread cooperative behaviour.

bad, values, and the like). However I am not making that leap: I am saying that it should not be made! I am comparing the structure of two arguments, and asserting that both are fallacious. The first argument was about the behaviour of a gas; the second argument was about the behaviour of human beings. The fact that the second argument is fallacious does not logically follow from the fact that the first argument is fallacious, just as the untruth of 'all elephants are mice' does not logically follow from the untruth of 'all prime numbers are even'. Rather, both arguments are wrong, and they are wrong owing to the same type of logical fault, namely the error known as non sequitur. They also have in common that in both cases the embodiment principle is being ignored.

Thus I readily admit that, by introducing moral categories, the above example steps well outside the area of discourse in which the previous chapters have been set. For this reason I will not discuss moral philosophy any further at this juncture, apart from what one must do implicitly just by writing about ideas of any kind. It is my purpose merely to show that the structure of scientific explanation is not completely at odds with the idea that moral philosophy is profoundly relevant to, and insightful of, human life. Far from it. Rather, the moral stature of human beings is a very natural extension to the structure of explanation that science offers.

Widely read works at the borderline between scholarly and popular literature have often promoted a view in which one scientific area (e.g. quantum gravity or Darwinian evolution or neurology) sweeps all before it and offers the whole of what deserves to be called 'explanatory' or 'objective'. This is a serious misrepresentation of science which should not pass unchallenged. Providing a more truthful picture is important in its own right, merely as a matter of sound scholarship. It may also help to prevent a fresh generation of scientists being brought up to think in naïve and ultimately inhumane ways.

I have called this a misrepresentation when it is done in a stark way which makes almost no attempt to grapple with other areas of discourse. However, essentially the same mistake may also be made in a less blunt way, a softer way which I would call a misunderstanding. This happens when a work does attempt to take different areas of discourse seriously, but in the end still asserts one over all the others. An example is the position taken by Sean Carroll in his *The Big Picture*, where he advocates what he calls a 'poetic naturalism'. Prof. Carroll intends to affirm that he welcomes and respects the fact that human experience and expression is multi-layered. However, what he actually says in the book is that the more complex or developed layers of discourse are merely 'convenient' human languages which do not convey objective truths. The 'poetic' for him is not prophetic, not a form of truth-speaking, but a form of entertainment, and that is all. By 'convenient', I am not sure what Prof. Carroll means; he could mean convenient for gaining power to manipulate the world, or convenient for stimulating the production of endorphins in human brains.

That may be his view; it may be the view of plenty of other people. I, personally, don't think it merits to be called a poetic view because it utterly devalues what poets do, but I have not been presenting arguments directly about that. The point that I have made in the opening chapters of the present book is simply that any view of scientific analysis which involves non sequiturs is a flawed view, and one such non sequitur is the widespread assumption that the whole cosmos is a machine in which notions such as moral judgement, artistic achievement, and social justice have no claim on objective truth, or have a lesser claim than the one made by our latest model of fundamental physics or evolutionary biology. Such an assumption is wrong. It follows that when there is a contradiction between our best grasp of a high-level and a low-level language, then it is not necessarily the case that the high-level is wrong and the low-level right; it could easily be the other way around. This does not invite wholesale rejection of well-established scientific ideas, but rather it invites openness to the possibility of further subtleties in any area of knowledge, and it respects the rich set of intellectual and other capacities that humans enjoy.

It is not my purpose here to present a full critique of Prof. Carroll's *The Big Picture*, but it seems to me that that work includes the above non sequitur as a major element, and therefore it does not construe science correctly. For this reason, it should not be called a 'scientific' view. Or even if it may, possibly, be allowed that his version of 'poetic naturalism' is a scientific view, it is certainly not the only scientific view. Another view, certainly a scientific view, is that all the areas of science grapple with what is so about the natural world, with equal aspirations to objectivity, and science does not invite us to commit category errors and non sequiturs. When Carroll writes, 'it's just the quantum wave function' (and everything else is a convenient way of talking) he commits the logical error discussed in Chapter 5 (the Babel fallacy). He should have written, 'it's a meeting of rocks and clouds and ants and trees and planets and people, with their wave functions, warts and all, whatever these things may be.'

There is a welcome gentleness in the modern wish not to be strident about large questions, but this gentleness should not be allowed to displace the sense of austerity which also accompanies truth. Both ingredients are needed in the political struggle of oppressed peoples. Oppressed people don't want to hear that their freedom and dignity is merely a convenient story that academic types can ponder in their armchairs. Are we going to deny their struggle a more objective validity? I hope not. In any case, my point here is that science does not say that we should.

By getting a better perspective on the structure of scientific explanation, we restore to rational discourse the ability to regard humans as whatever humans truly are. Science does not invite us to deny that we are persons, alive and conscious. When science is seen correctly, then to be insightful about science does not involve or invite disparaging remarks about the humanities, and it allows that there may be high-level languages expressed in human life, whose categories cannot be adequately expressed in terms of lower-level concepts. The point is not that I now claim to know for sure what humans are; the point is that I claim *not* to know, and that no one can know with anything but a tentative knowledge which draws on all aspects of experience.

This point of view restores humanity in the following sense. What I have offered is a way to see that each human person is not just a complex collection of chemical reactions, because that way of speaking implies an untruth. It implies that the speaker already knows what sort of thing a human-shaped collection of chemical reactions is. But that is wrong, because the only way to discover what human-shaped collections of chemical reactions are, is to meet with them. Only then will one discover the all-important shaping principles, and now of course we are talking about principles deeper still than those which find embodiment in monkeys. We are talking about values, such as the priority of forgiveness over revenge, and of teaching over dictating, and of love over hate.

These are not meaningless offshoots of some sort of genetic struggle that happen to promote survival, nor are they charming ideas which just randomly pop up in machines such as brains. No. These are examples of the beautiful symmetry principles that can find expression in human behaviour. It is not our job as scientists to deny this. It is our job to live up to it.

Reflection

Vector

When I was young I saw it all at once:
a finger, a crayon, a twig, a railing, a line.
I traced it on the sides of things,
and held it in my hand. It marshalled
me the way that I was going,
and an instrument to use.
A friend, and fun, but it could also
sting from nettles and splinter skin.

At school we got it down and
Descarted it into x and y and z.
I watched it spin in mind-space,
for hours, and days, and years.
Sometimes z stood still
and x and y joined hands to dance.
When one bowed low the other reached.
Then one burrowed, and hung, a bat.
The other followed, and then begin again.
Sometimes z announced it would be x,
and y, z and x, y, and one learned
to make sense of this,
from a certain point of view.

But lately I have begun to find it whole again:
the thing in itself.
The dagger in the mind.
It has become not letters but a word,
a cog in the vocabulary of thought,
which is the way I think and even see.
When someone asks,
'How do we know that x can be y,
and y, z, and z, x?'
I can only try to begin again, and show
a finger, a crayon, a twig, a railing, a line,
and trace it on the sides of things,
and show it on my hands.

7

Purpose and Cause

> The distinction between cause ('owing to what?') and purpose ('in order that what?') is spelled out. I unpick the confusion of these concepts that takes place in the writing of Richard Dawkins, especially in *The Selfish Gene*. Intellectual discipline must be respected; the question of purpose is informed but not answered by science. Furthermore, high-level languages applied to human life are valid and insightful. The products of Darwinian evolution are not defined by their role in the propagation of genes.

Two friends are standing on top of a mountain, looking around them. One notices that the mountainsides are covered in looming cliffs and yawning chasms. It seems impossible that anyone could ever have climbed up.

'How did we ever get here?' he asks.

The other, who is a bearded gentleman who taps people on the shoulder before replying, then points out that, weaving among all the cliffs and chasms, there is in fact a gradually sloping path. Having guided his friend down this path to the mountain bottom, and back up again, he rests content. The first then asks his next question:

'Thank you. Now could you tell me where we are?'

The bearded gentleman replies, 'I have no idea. We are on top of some mountain or other, I think.'

The previous chapter concludes the development of ideas underpinning the first main part of the book. The next few chapters continue the theme of the structure of scientific explanation, by looking in more detail at natural history and the way living things have come to be as they are. The aim is to promote an accurate overall picture of that process, and to unpick some deep-seated misconceptions that circulate in the popular mind, and in some professional circles. The main such misconception that I want to unpick is the one that supposes that the principal function or goal of each living thing is the proliferation of its DNA. A professional philosopher might feel that I am attacking a straw man here, but I have grown tired of reading pieces of literature that present themselves as 'explaining science' and proceed on the basis of assuming this very thing. A particular *bête noire* of mine is the claim that forests are 'poorly designed' or 'inefficient' because the trees are 'unnecessarily' tall (you see, DNA could have been passed on by smaller trees if all trees had been smaller.) People who say things like this tend to do so with a kind of smug assumption that they have 'seen through' the misleading appearance of things (the appearance that might make one feel that trees are majestic, for example) and found 'the truth', by their application of 'scientific' reasoning. If one thinks that the aim or purpose of having a forest is to make more copies of certain molecules (namely, tree DNA) then one concludes that there is a great waste of resources. However, why would one ascribe that aim to the observed process? Such a statement is not a logical deduction from the role of DNA in the physical causation, because the notion of *purpose* is entirely unrelated, in logic, to the notion of *physical cause*. The purpose or most enduring role or significance of a thing is either not logically connected to the physical causes which gave rise to it, or else the connection is much less direct.

This is the central issue that the present chapter is about. The significance and role or purpose of a complex thing is not indicated by what happens to the smaller, simpler things of which it is made, nor is it indicated by the physical process which shaped or built the complex thing. Rather, the role is indicated by how that complex thing eventually contributes at all the levels of discourse in which it can operate.

For example, a mammal is not merely, nor primarily, a bag for carrying water around, even though a mammal is mostly made of water. Rather, a mammal is a being capable of much more profound contributions such as perseverance and even, to some extent, empathy. Also, a shaping or building process does not unambiguously indicate the wider purpose of what results. If one was ignorant about this, then one might conclude, from observing the training of a musician, that the purpose of a musician is to play scales accurately. Or, the production process for aluminium foil might make one conclude that the chief role of aluminium foil is to form a tightly rolled cylinder that can fit in a box. One could multiply examples taken from human life, but such examples involve choices and plans made by conscious agents, whereas the process of biological evolution does not,* so there the whole question of role and purpose has to be considered afresh. Some people will go so far as to make round pronouncements along the lines that no natural phenomenon has any purpose or enduring significance (such people tend to be silent as to whether their pronouncement is itself a natural phenomenon). To me, such a pronouncement is utterly lacking in understanding and seems to sever the very humanity which ought to underlie all our attempts at science. I think it is really a reaction against a too-hasty reading of purpose into natural phenomena, which can become an obstruction to understanding them properly. But just because we do not possess the means to make a confident judgement about purpose, it does not follow that no purpose exists.

This chapter addresses the distinction between cause and purpose, and the fact that this distinction must be carefully maintained when we undertake scientific investigation. This is no less so in the case of Darwinian evolution than it is in all other areas of science.

Here, the word *cause* is in the sense of cause and effect, and identifying good answers to the question 'owing to what ...?'. *Purpose* is in the sense of goal or end, and identifying good answers to the question 'in order that what ...?'. It is basic, and obvious, that these are different questions that address different types of issue.

The notion of 'cause' is the more philosophically straightforward of the two, though it does have some subtlety. Roughly speaking, if a circumstance B would have been changed in some way if a prior circumstance A had been other than it was, other things being equal, and if A involves things that are located near to B, and one can trace or reasonably establish a connection from A to B (as opposed to a common cause of A and B), then there is a *prima facie* case that A was among the causes of B. Or if there is a universal law according to which circumstances like A are always followed by circumstances like B (e.g. A = 'hammer hits nail'; B = 'nail moves away from hammer') then A causes B.[†]

---

* I mean the process set out as a natural process, within the domain of competence of ordinary science; that is all I wish to discuss here.
† There are philosophical positions which amount to a denial of the notion of cause and effect. One might regard events in space and time as no more directly causing one another than do the points of paint in a pointillist artwork. I am inclined to reject that and to assert that the natural world does have the dignity of causality, but this is a conferred dignity.

The notion of 'purpose' is more subtle because goals and ends are qualities that are associated with thought and intention, and there is no straightforward sense in which any physical object carries a uniquely identifiable purpose in and of itself. However, the question of purpose is an important question, and often a defining one. The verb 'to be' is often connected, in language, with the notion of purpose. If one examines dictionary definitions of human artefacts such as tools and everyday objects (chairs, clothes, pens) then often the use to which the object is put features in the definition, as in 'a chair *is* a solid object suitably proportioned to be comfortably sat upon' or 'a pen *is* an instrument suitable for writing.'

When we carry out scientific analysis, we learn much about cause–effect relationships, and this helps us to appreciate the role that various physical objects play in the grand scheme of things. Sometimes people follow a train of thought that runs roughly as follows:

1. Things like A almost always and almost everywhere lead to things like B.
2. This is the essence of what A does. It amounts to a defining statement: A is that which causes B.
3. The purpose of A is to cause B.

For example,

1. Pens, broadly speaking, are solid objects containing ink and used by people for writing.
2. This is the essence of what pens do. It amounts to a defining statement: a pen is that which is used for writing.
3. The purpose of a pen is to facilitate writing.

However, the steps in this sequence are not logical inferences. They are judgements made by an assessing individual. Consider the following example:

1. Pens, broadly speaking, are solid objects suited to be held in the human hand and which, when dragged across paper so as to leave a trail of ink, give exercise to the muscles and coordination of the hand.
2. This is the essence of what pens do. It amounts to a defining statement: a pen is that which exercises the human hand.
3. The purpose of a pen is to serve as an aid to muscular exercise.

This illustrates the fact that a sequence of this kind is not a logical sequence, and any statements about purpose that are arrived at this way can be legitimately questioned.

Here is a slightly longer example, in the form of a short fairy tale about life on Earth.

The Imperialist Water

A long time ago, you could only find water in a few places: the ocean, rivers and streams, and clouds. However, some water molecules, after going through the cycle of evaporation, condensation, and precipitation a few times, got ideas of colonizing the land and the air. So, with this programme in mind, they started to gather themselves around various floating amino acids and proteins, and they arranged for some chemistry in aqueous solution to take place. Slowly, more and more complicated sequences of chemical reactions were going on, and the water molecules by chance hit on a sequence which constructed little bags for the water to go in. These bags were themselves floating in water, but it was a start.

As the imperialist water molecules really wanted to dominate the land and the sky, they kept up their steady work, and they managed to hit on a way to attain exponential growth of the structures they were making. They co-opted some genetic machinery to do the work for them. Eventually they made multicellular organisms, and, slowly, they developed ways to allow these organisms to extract oxygen from water and even from air. It all happened by slow accumulation of change; the structures that did a better job of surviving and reproducing tended to proliferate, and consequently did more of the work of transporting water.

Eventually the imperialist water molecules got themselves up out of the ocean and started to colonize the land. They created plants and animals, their slaves, and thus achieved their purpose of propelling themselves all over the planet. What weird engines of self-propulsion would the millennia bring forth? Birds and monkeys, elephants and beetles. A monkey is a machine that carries water in the trees, a bird is a machine that carries water in the air.

And human beings too, of course. We are machines built by water whose purpose is to make more machines able to carry water. Flowers are for the same thing as everything else in the living kingdoms, for spreading water about. That is exactly what we are for. We are machines for transporting water, and the transport of water is a self-sustaining process. It is every living object's sole reason for living.

We are water carriers—robot vehicles blindly programmed to carry the imperialist molecules known as water. This is a truth which still fills me with astonishment. Though I have known it for years, I never seem to get fully used to it.

The first example I gave above—the example of the pen and its uses—is there simply to illustrate the fact that cause and effect are not always a sure guide to purpose, in the case of artefacts designed and used by humans. I don't deny that a more thorough study of pens, including the way they are presented in advertising campaigns, for example, would arrive at the conclusion that to facilitate writing is a better assessment of their purpose in human culture than to exercise hands. In fact, the role of wider culture here is a helpful indicator of how purpose is signalled not just by what objects do when they are used, but also by how people think about them and what further role is played by the effects caused by the objects.

The second example—the story of the water molecules—is not intended to deny that the genetic apparatus has a lot more to teach us about the detailed structure of living things than does water, but the story does serve to illustrate the nature of the question at issue when we consider function and purpose. In order to fill in values for x and y in a phrase such as 'we are x whose purpose is to y', it is not sufficient to study the sequence of physical cause and effect that led up to there being such things as ourselves. Nor is it correct to say, '*from the point of view of water molecules*, we are . . . whose purpose is to . . .'. That is not correct because who knows what water molecules would want? Maybe they wanted to stay in the ocean and the story has been an unmitigated disaster for them! Of course this merely underlines that metaphors have limited uses and some succeed better than others. The point is that if anyone introduces a metaphor, then they can use it for illustrative purposes only. They have to justify by other means whatever conclusions they intend the metaphor to illustrate.

I am not going to discuss further the notion of purpose at this stage. The above serves as a preamble to this chapter, whose subject is the consideration of purpose and cause in the context of evolutionary biology. The chapter is intended to help the general scientific reader get clarity about this. I will not be saying anything controversial, and I will not need to address the detailed considerations which the professional scientific community brings to bear in this area. My aim is simply to help undo the confusion which was caused by Dawkins's famous book, *The Selfish Gene*,[24] and others like it, in which purpose and cause are muddled, as I will explain.

7.1 Why Has an Anteater Got a Long, Sticky Tongue?

Here are a variety of possible answers to this tongue question. The list includes answers with which I have varying degrees of sympathy, and it includes some which I think are sufficiently bad answers as to be essentially wrong:

A. Because that allows the anteater to extract ants from long, narrow holes.
B. It is owing to the fact that a set of genes for tongue length, and for tongue stickiness, went through many cycles of variation and reproduction, and those that gave rise to tongues suited to catching ants promoted the health and hence reproduction of the animals with them, and thus were amplified in the population of a species that also had other adaptations suited to finding and digesting ants.
C. It is a by-product of a process whose purpose is that there be more genes for long, sticky tongues in the world.
D. It is a result of a process whose purpose is that there be more long, sticky tongues in the world.

Figure 7.1. Giant anteater *Myrmecophaga tridactyla*.
Source: Meyers Konversations-Lexikon 1897; shutterstock 97038395 Hein Nouwens.

E. It is a by-product of a process whose purpose (or goal or chief result) is that there be more self-replicating molecules in the world.
F. It is part of a process whose purpose is that there be a variety of living entities able to negotiate their lives in a variety of degrees of complexity and existential depth.
G. If by 'why?' is meant 'to what end?' then the question is meaningless. It commits a category error by referring to something (possession of a tongue) as if it had a property or quality which it cannot exhibit (namely, purpose). It is like asking about the purpose of the outcome of a roll of a dice.
H. It is in order that the anteater will promote (through its own health and reproduction) the proliferation of genes for long, sticky tongues.
I. It is owing to inheritance, and to the fact that anteater ancestors fared better when their genes acquired changes that resulted in more nearly optimal tongue length and stickiness (optimal for promoting anteater reproduction, that is). With a tongue that is well-suited to extracting ants, a member of an ant-eating species (i.e. one that also has other adaptations suited to finding and digesting ants) will promote, through its own health and reproduction, the proliferation of genes for such tongues.
J. Because if it had a short, slippery tongue it would not be an anteater.

Note that, owing to an ambiguity in the English language, the same word 'why?' can mean 'owing to what cause?' and also 'in order that what purpose be accomplished?'. The word can also have other meanings, but these two are the main ones that people ordinarily have in mind, and both contribute to the notion of 'explanation'. In the above list I have provided answers which address both types of question.

And there are further distinctions to be made. The question 'in order that what?' can mean 'what was the goal aimed at by a consciously reflective agent?' or it can mean 'what function does it serve in the given context?'. For example, the legs of an ant have a function in the life of an ant, which is to provide a means whereby the ant can move itself around, but it is not clear whether any purposive and capable agent so arranged things that these insects would have the legs they have. In the context of scientific study it is usual to rule out the latter kind of consideration, not as impossible or meaningless, but simply as not relevant to the study being undertaken.

One can make a comparison with the study of a machine such as a steam engine. It is one thing to ask what the function of a piston is in the working of the engine, it is another to ask whether the whole engine has been made by someone with some conscious intention, for example the intention to undertake a journey or provide transport. When someone says 'the purpose of the piston is to convert steam pressure to motive power', this may be a way of saying 'the function of the piston in the engine's working is that it converts steam pressure to motive power, but I know nothing of whether the whole machine is going anywhere desirable or desired'. On the other hand, the same statement about the purpose of the piston might be a way of saying 'the goal is to go on the journey, and the purpose of the piston is to play its part in making this possible, that part being conversion of steam pressure to motive power'.

Now, when studying the natural order of things, it is clear that one is not in the same position as studying a human artefact such as a steam engine, and it is also clear that one is studying something fascinating, marvellous, and very complicated. For all these and other reasons, it is the standard practice in science (biology, ethology, zoology for example) to restrict study and arguments to finding causes within the framework of natural processes, and elucidating functions within those processes and at a modest scale. One hesitates to speak of the function of the whole biosphere, for example. This restriction is useful, but it is not the same thing as saying there is no larger purpose to the natural world. Rather, it is a choice to remain neutral or silent on that; to leave it open till after the scientific study has done what work it can to help figure out what is going on.

Now let's consider the answers A–J listed above. Answer A might be called 'the child's answer' and I have some sympathy with it. It is not completely wrong, because it asserts a connection between the properties and use of the tongue, and there is such a connection. This answer asserts, mainly, that the properties (long, sticky) are suited to the use that the anteater makes of its tongue (extracting ants). However, it is not a very full answer because one naturally would like to know more about it, such as answers to the questions, 'by what sequence of cause and effect did this state of affairs come about?' and 'to what end, if any, did this state of affairs come about?'.

Answer B is a brief presentation of standard components of evolutionary biology, and it is also a good answer, in my opinion. It gathers up answer A (because it notes the connection to properties and use) and it briefly sketches the sequence of cause and effect whereby the situation has come about.

Answers C and D each act as a prod, a thought-provoking throw-away line, almost a joke. Neither warrants to be taken very seriously, but if it is correct to say that there is purpose being worked out in the world (and I think it is), then I am inclined to say that a long, sticky tongue has more of the nature of the kind of thing that might be an end, rather than a means to an end, than does a length of DNA that codes for tongue properties. I will return to this in what follows.

Answers E and F further extend the rumination on purpose. I would say that E is flawed in comparison to F, for a reason similar to the one just described. That is, an ecosystem with creatures able to display existential depth has more of the qualities that one associates with a purpose or end, as opposed to a means to an end, than does the proliferation of DNA molecules. The phrase 'existential depth' is a reference to the range and type of qualities that an entity can experience and express.

Answer G responds by denying the validity of all responses that relate to purpose. This is, as a matter of logic, an answer that might possibly be correct, but I want to alert the reader at this point to the fact that this answer does not have greater support, from scientific study, than answer F. In fact, *the intellectual discipline called science cannot adjudicate between G and F*. In practice each of us forms an opinion or tentative conclusion or working assumption on issues relating to purpose and meaning. Scientific study can help us to elaborate or set out more fully what is our opinion on G and F, but it does not provide that opinion. I already alluded to this fact and I will return to it.

Answers H and I present statements that are similar to one another, but one is about purpose, the other is about cause. Insofar as they make any sense, they do so only if one has the Darwinian process already in mind. Once one places these answers in that context, then H is dubious and I is right. Answer I amounts to a repeat or an extension of answer B. It extends that answer by giving some more of the relevant background.

Answer J is the kind of answer that a friendly adult might throw back at a child, to make them laugh and also make them think. It has in it the seeds of something quite profound.

In his celebrated 'gene's-eye view' of evolution, presented in *The Selfish Gene*, Richard Dawkins has advocated a position which includes, roughly speaking, all the above answers except F, and which emphasizes and re-emphasizes E again and again:

'What weird engines of self-preservation would the millennia bring forth?';
'A monkey is a machine that preserves genes up trees, a fish is a machine that preserves genes in the water';
'their preservation is the ultimate rationale for our existence';
'we are their survival machines';
'We are survival machines—robot vehicles blindly programmed to preserve the selfish molecules known as genes. This is a truth which still fills me with astonishment. Though I have known it for years, I never seem to get fully used to it.'

Note the phrase 'the ultimate rationale' here. In their context, the use of the verb *to be* in the above statements has that sense: 'a monkey is . . .', 'a fish is . . .', 'we are . . .', are intended to be understood as statements about what these animals are first and foremost, or in essence.

These statements function in *The Selfish Gene* as comments on evolutionary biology. They are intended to help the reader get a good understanding. However, the book conflates and muddles the notions of *cause* and *purpose*. The distinction between these notions is never got clearly in view. Instead, the book carries the reader along in an argument that is largely about cause and sometimes about function, and then abruptly presents summary statements as if the argument really referred to purpose. This is done without justification, and in fact it is not justified.

In this way Dawkins presents at the same time a large number of fascinating and important insights into evolutionary biology, and also an overarching interpretation or opinion of what the whole story declares. One does not need to accept that interpretation, however. One can have all the evolutionary biology while being of the opinion that F is a more coherent statement than E, for example, because F deals with the whole concept and category of 'purpose' in a way that is more suited to that concept and category.

This deserves a further comment. I find myself intrigued by the fact that, when one reads *The Selfish Gene*, one finds oneself carried along by the argument, to such an extent that one begins to feel as if E could possibly make sense. I think this may be because the 'gene's-eye view' is there presented with very liberal use of metaphor, so that genes, or more generally, replicating molecules, are regularly and repeatedly attributed with colourful planning and scheming abilities; they are 'past masters of the survival arts', 'manipulating it [the outside world] by remote control'; 'they created us'; 'Of course they march on. That is their business'; 'the tyranny of the selfish replicators'; 'the replicators have built a vast range of machines'; 'Each selfish gene therefore has its loyalties divided between different bodies'; 'it is the business of genes to program brains in advance'; and so on. In all these phrases, and many more, it is simply assumed that replicating molecules are agents to which categories such as 'past master', 'the survival arts', 'manipulate', 'create', 'their business', 'tyranny', 'selfish', 'exploit', 'loyalties', and 'program brains' apply. To be precise, this is not assumed but rather invoked as an imaginative aid to understanding. I have no quarrel with imaginative aids to understanding when they are used to promote correct conclusions. But something else is going on here. Each example, taken alone, can be received as a thought-provoking metaphor, but when the metaphors are repeated and rehearsed so forcefully, and when they all represent versions or extensions of one overarching metaphor—that of selfish scheming to replicate—then something happens in the mind of the unreflective reader. The scheming-replicator metaphor begins to take on a life of its own, so that the whole idea that a gene cannot actually have a view, that there is, in fact, no gene's eye, begins to be forgotten. Thus the simple truth that short strands of DNA cannot scheme and manipulate is not completely forgotten, but its emotional power fades; instead, the emotional power of the idea that genes are manipulative agents on a campaign of unrestrained self-cloning takes hold and the supposed 'gene's-eye view' becomes the view held by a human being. But the human being balks at this; one senses that one has been misled somewhere, but one is not quite sure where. I think this tends to leave one in the position of feeling that the whole story of life on Earth is basically meaningless. One retains the excitement of the idea that one has, in the 'selfish gene', a guiding metaphor that will forever yield insight into the results of Darwinian evolutionary processes, and this is indeed a rather exciting idea. But at the same time one senses that the notion of purpose has evaporated, or been replaced by something thoroughly banal.

However, the notion that there is no meaning, or that the meaning is something banal such as the production of copies of molecules, has not been argued. It has been assumed.

Perhaps, at this point, the reader may object that what *The Selfish Gene* is arguing is not really about the notion of 'to what end?'; it is only concerned with 'owing to what?'. Dawkins does sometimes refer directly to the *purpose* of things like flowers, or invokes the notion of 'what they are for', but one might wish to say that this is just a hyperbola, and really his intention is to press home his point about the nature of the Darwinian process as a sequence of cause and effect. In this case, a phrase such as 'the ultimate rationale for our existence' means something like 'the most significant item in the physical causes which resulted in our existence'. The claim is then that the facts known as Darwinian (i.e. exponential proliferation of genes by reproduction, in the context of restricted resources and the associated competition) together constitute the central component or controlling idea of the physical sequence which led up to the existence of human beings (along with other living things). Such a claim is debatable because it fails to mention the shape of the abstract space which is explored by the Darwinian mechanism: the point I already discussed in Chapter 4.

From a chemical point of view, one might say that the 'ultimate rationale for our existence' is chemical reactions and energy dissipation. People don't generally assert that because they can see that it is missing the point. When looking into the causes of our existence one wants to know, in particular, what it is that we are saying when we notice that a human being is different from a spider, or a bear, or a tree, etc. Mentioning that both are outcomes of chemical reactions doesn't help very much. Mentioning that both are outcomes of inheritance and variation and selective pressure helps a bit more, but still not very much, because it says close to nothing about the structure of the high-level patterns that constrain the results that can happen. If you doubt that there are any such high-level patterns, look around you (and reread Chapter 4).

I think it is fair to say that there is a muddle and conflation of cause and purpose in much of Dawkins's writing about living things. This may be because it is indeed quite difficult to tease these two concepts apart in the case of Darwinian evolutionary biology. The situation is different from that of simple physical ideas, for example, where one rarely if ever finds it appropriate to invoke the category of 'purpose'. We don't usually find it useful, not even useful to the imagination, to say that the *purpose* of electric charge is to source the electric field, for example. Statements of that kind are either incoherent, or else metaphors that do not succeed. It is better, in the sense of tending to greater clarity of thought, if one just says that the charge sources the field, and leave it at that. This is not to rule out of valid discourse all statements of purpose; it is rather to postpone them and to recognize, correctly, that they address a different category of issue to statements of cause and effect.

The confusion in *The Selfish Gene* arises because there is something about the feedback involved in the whole business of replication and inheritance that makes cause, function, and purpose less easy to separate. However, that does not mean one should give up the attempt to tease them apart in order to think clearly.

7.1.1 Premature Teleology

As science has progressed over several hundred years, it has become more and more clear that one should be cautious of *premature teleology*. That is, one should be cautious of declaring too soon that one knows about what *purpose* a given part of the natural world serves. Whereas it is largely unproblematic to declare 'this is thus and so *owing to* . . .', one should be shy of declaring, 'this is thus and so *in order that* . . .'. For example, one may assert:

> K. The lioness is protective of her cub, and this protective behaviour is in part owing to the fact that it is promoted by the very genes which thus come to be more common in the gene pool.

However, one may not assert, without a further and careful justification:

> L. Insofar as any lioness has a purpose in the grand scheme of things, its purpose is to promote the production of copies of genes like its own.

The central point here is that *the second assertion does not follow from the first*. There is no implication, in logic, from a statement of cause and effect to a statement of purpose. The discussion in *The Selfish Gene* gives useful insight into the first assertion above, but it then leaps to the second, illogically and with no case made.

Let us use the label 'SGA' to refer to a person who accepts the position argued in *The Selfish Gene*. That is to say, SGA refers to one who accepts both the assertions about cause and the assertions about purpose that are made there. Such an SGA need not think that there is a logical inference from one to the other type of assertion (SGA may or may not be aware that such an inference would not be logical).

Now suppose someone were to assert (as many children do assert) that they do not know the purpose of flowers, but they guess it may be to make the countryside pretty. I guess that SGA would wish gently to persuade such a person that the true purpose of flowers is to promote the replication of flower-genes. SGA will say that the intuition about beauty was misleading, and, from a scientific point of view, incoherent, whereas SGA's view is the scientific one.

This is nonsense. In order to earn the adjective 'scientific' an account must respect logic and not bring in illogical inferences from cause to purpose by means of heavy use of metaphorical language.

This is not to suggest that a child of any age shouldn't be shown the fascinating journey of inheritance, variation, and natural selection, including the point that the process acts at the level of the gene pool, and that a fish is to some degree a machine that preserves genes in the water, and a flower a machine that preserves genes in the countryside. However that same conversation should also respect and indeed champion the right of the child, or anyone else, to come to a rounded view about what purpose all this might be achieving.

Let us visit in our imagination the Rivonia Trial, at which Nelson Mandela and his co-defendants were on trial for their lives on a charge of acts of sabotage against the state of South Africa. Suppose that defence attorney Bram Fischer had offered as an argument something like the following: 'The court must act in accordance with what is strictly true and scientifically established. My client Mr Mandela, and his co-defendants, are robot vehicles blindly programmed to preserve the selfish molecules known as genes. I also am such a blindly programmed robot. So are we all.' If Mr Fischer had said that, then he would either have been making a joke in poor taste or else he would have been thoroughly foolish, or else he would have been saying one thing and meaning another. Such a statement does not deserve to be called 'scientific', let alone 'scientifically established', because it fails to take even a little effort to consider what kinds of categories might be appropriate when speaking of humans so as to express what manner of thing a human essentially is. The structure of science is not one which dismisses the higher-level categories and 'explains them away'. That is not what 'scientific' means.

In the previous paragraph I have translated a statement or a sentiment into a context where its author might not have used it, and in the interests of fairness I wish to acknowledge that that is so. The point of the illustration is to show that in a context where careful, true, and insightful speech about human beings is wanted, the statement about genes is almost entirely irrelevant. Far from being a centrally important insight or defining clause, it can be completely set aside, with no harm to the business at hand.

Premature teleology tends to result in a mindset that stands in judgement on the natural world before it has been fully understood. For example, if one thinks that the purpose of a tree is to reproduce genes like its own, then one will consider that any lack of efficiency in achieving this end is just so much wasted effort. Owing to competition for sunlight, in a forest the taller trees tend to do better, and over many generations this results in a forest in which every tree is tall, whereas if all were shorter then all would make a great saving in materials and time (Figure 7.2). So if the purpose of trees is their own gene survival, then there is a great deal of 'waste' in the process. But if one avoids premature teleology then one is free to say that one does not already

Figure 7.2. A forest, not a wasteland.
Source: Floris Slooff; shutterstock 9212323.

know what purpose trees serve, and one is open to seeing things more in the round. One can take into account that the bark itself supports another ecosystem, and the trunks are climbed by insects, and the shade is inhabited by other animals. Perhaps in fact very little or nothing is wasted in this example.

Indeed, even without all this, merely to be tall, where height is difficult to attain, is something of an engineering achievement, and deserves recognition as such.

7.2 Science and Intellectual Discipline

A reader who is shy of accepting the argument of the present chapter, perhaps because they find it obvious and wonder what I am trying to achieve, or because they suspect I am misrepresenting a position or attacking a straw man or worse, should consider the fact that it has become common for people to pronounce judgement in the way I alluded to above, by claiming for example that forests are hugely wasteful, or by suggesting that the sequence of events known as an 'evolutionary arms race' has the kind of questionable morality that aggressive and suspicious human behaviour has. In the interests of the public understanding of science, we must point out that such inferences are illogical and unjustified.

This is not to say that it is never appropriate to say that some sort of defect is present in the natural world, or that something like the starvation of a young polar bear is a great sadness and has at least some of the flavour of an injustice. However, a teleology which asserts that such a polar bear has altogether missed its own purpose seems to me premature. I think that even in its short life, such a polar bear has expressed much that is valuable and therefore it has given its efforts to some purpose in the grand scheme of things.

Now I will reiterate a point I made after displaying the list of answers A–J. It is the point about the nature of the activity we call science. Someone might reply to this chapter (and I am indebted to N. Carl and C. Jacobs for putting this reply squarely into words):

> Indeed, you are quite right that purpose and cause should not be confused—and perhaps Dawkins does make this mistake. In fact, the notion of purpose has no place at all in science! Just as it is not the purpose of the Sun's mass to make the planets go around her, it is also not the purpose of a gene to promote itself in the gene pool. This is exactly what

the Darwinian theory of evolution is about: evolution is blind, not guided or part of a design, and as such it has no goal or purpose. It's just a (more or less) random sequence of mutations and environmental influences.

I go along with much of this statement, but not the last sentence and a half. Evolution *is* guided, by the truths of existence, including, for example, the tautologies listed in Section 4.3. That aspect is what I have discussed in previous chapters. In the present chapter I am engaged in questioning the unsupported use of the word 'just' in the phrase 'it is just a . . .'. Such unsupported usage both fails to appreciate the embodiment principle (Chapter 4) and commits the Babel fallacy (Chapter 5), and the present chapter exposes a further related fallacy. I am saying that the inference from mechanistic model to lack of purpose is not justified. It is not a logical inference: it simply does not follow. In the words of Monty Python, such an inference is not 'resting' or 'stunned' or 'pining for the fjords'; it is a dead inference; it has ceased to be. If it had not been nailed to the perch then it would be pushing up the daisies.

Let us look into this a little more closely. It is quite possible to address oneself to the following job: 'Let us deliberately set to one side the whole notion of purpose or end in what goes on in the phenomena we are studying, and just see what we can find out about the phenomena under that decision; that is, in that sort of intellectual exercise.' One notes that this is the sort of choice made in either all or almost all scientific study. Indeed (as you say) it is almost a definition of science (but I won't get into that). When one makes that choice, one discovers much. One discovers that a great deal of sense can be made. But it is utterly inappropriate to then suppose, at the end of such a study, that one's deductions *imply* that the phenomena had no purpose. Rather, the judgement about purpose, or the lack of it, is still before one. It has not been decided but postponed.

I am not here clamouring for the correctness of whatever judgement I might make about the purpose of things; I am clamouring for and indeed *insisting on* the fact that an assertion such as 'it has no goal or purpose' is a religious response not a scientific deduction. That is, such an assertion is not one that is handed to us by the data, it is one that we generate out of our sensitivity to meaning or the lack of it in the data.

Obviously an impersonal physical process or set of phenomena does not have anything like awareness, so it does not have purpose in the sense of an end that it, the process, consciously intended, but that is not the point here. The point is, does the process or phenomenon serve some end that we can perceive? The scientific study of the phenomenon itself does not answer this question. Such study does not imply that phenomena have purpose, and nor does it imply that phenomena have no purpose.

Scientific study informs a judgement about purpose (about the question, 'in order that what?'), but such study does not, and indeed *cannot*, make the judgement, because it is the very judgement that the methods of experimentation and model-making do not and cannot explore. It is the very thing that the set of intellectual tools called *the scientific method* has no power to address, or to come to any conclusion on, one way or the other.

Once again, this silence—this refusal to pronounce—does not mean, and should not be interpreted to mean, that the 'scientific' view 'is really' that there is no purpose. To assert that in the name of science is not to respect science but to misunderstand or abuse it.

7.3 The Multi-Layered View

Now let us return to statement (K) (page 87). There is a cause–effect relationship between the Darwinian mechanism and the instincts of a lioness, but it is far from the whole truth. First, there are further cause–effect relationships that one could mention, such as the increase of entropy which is involved at every stage, and the electromagnetic forces which bind molecules, and so on. Second, and this is more pertinent to the present discussion, the connection between the Darwinian mechanism and the lioness's behaviour is mediated via a rich combination of further structures such as cell walls, oxygen metabolism, eyes, livers, mating rituals, cognitive powers, recognition of likeness, and so on, and *the connection involves that these structures all satisfy principles of physics, chemistry, engineering, social relations, how learning happens, what sympathy is, etc., which are nothing to do with Darwinian evolution* per se. These significant contributory principles are not invented by evolution; they are truths which evolution cannot but respect (recall that you can't evolve negative mass, etc., the point discussed in Section 4.2).

All living things are the result of a specific, rather than an arbitrary, kind of trial and error. It is one that tries out ways of being alive and gradually discovers greater richness that existence can furnish.

With this in mind, let us extend and improve our assessment of lioness behaviour by replacing statement K with the following:

M. The lioness is protective of her cub, and this protective behaviour comes naturally, without the need for much exercise of a will, because evolutionary processes have furnished the lioness with a brain and instincts that incline her to recognize the cub as an object of her protective instincts. The way inheritance and replication work tends to result in this sort of instinct arising more widely insofar as the object of protection has the same genes as give rise to this behaviour in the protector.

This more extended statement is an improvement on K because it correctly recognizes and duly respects the fact that what the lioness does is *recognize her cub as an object of her protection, and behave accordingly*. This is not a mirage, nor an empty statement. It is a correct statement about cognitive abilities and social relations in lions.

Just as it is correct to say that cathedrals have arches as well as stones, it is equally correct to say that mammals are capable of exhibiting genuinely caring behaviour. The stones do not explain away the arches, and the genes do not explain away the caring. The caring is there, and whatever 'selfish' or 'eager' character genes might have has been subsumed into something at another level of structure and of explanation—something quite beautiful in this instance. Most (very nearly all) commentators will agree on this, but it is liable to be forgotten or sidelined when the discussion makes genetics and the mechanics of the Darwinian process central.

In M we still have not said anything about purpose. Who knows what the purpose of a lioness is? I already alluded to the fact that science is an academic discipline in which we mostly avoid addressing this question. We fight shy of it, because we are not sure how to assess whether one answer or another is better. Someone studying social life among lions can say where the mothers and the hunters fit in; someone studying ecosystems on the African plain can describe the role of predators; someone studying human behaviour can assert that humans find their mammalian cousins rather wonderful. All these are statements about how a lioness functions in one wider system or another. All hint at notions of purpose; none provide a clear sense of overall purpose.

It is possible to get by in human life without answering the question of purpose. One can even shrug one's shoulders and say perhaps there is no purpose to the grand scheme of the natural world, or that the only purposes are the ones that humans and other animals invent for themselves. This is all possible, but it seems to me to go against the evidence—the hints that I just alluded to. It seems to me that this great engine of the natural world is not just going through the motions; it is going somewhere. So I am ready to suggest that ordinary animals, and even inanimate things, can be said to have purpose independent of any projects or wishes that I or the rest of the human race may have. I fully admit that none of us is adequately gifted to pronounce on questions of this kind with confidence, but I do feel that we are called upon to champion the notion of hope over fatalism and disdain. I feel that for a human to pronounce the judgement that the natural world has no purpose is a more pompous thing for a human to do than for them to say that they suspect the natural world has purpose and they are willing to act as if this were so. But the recognition of purpose in nature has to be combined with a great sympathy for what nature is.

Perhaps the purpose of a lioness is to dissipate entropy as quickly as possible (I doubt it). Perhaps the purpose of a lioness is to produce more lioness genes (I doubt it). Perhaps the purpose of a lioness is to kill, eat, and copulate (I doubt that too). Perhaps the purpose of a lioness is to be a lioness (this seems to me to be on the right track). Perhaps the purpose of a lioness is to express whatever good a lioness is capable of expressing (I am inclined to pick this one).

Perhaps the purpose of a lioness is to express an aspect of what the verb *to be* can signify.

I will finish with a comment on the structure of scientific explanation, which has been in the background of some of the material in this chapter.

Whether or not one has in mind a 'gene's-eye view' of evolution, one must admit that genes themselves are a valid concept with an explanatory role. This can fit into the structure shown in Figure 3.2 as long as one throws out the incoherent use of the word 'purpose', as I have explained. However, if one tries to fit the gene's-eye view into the structure shown in Figure 3.1, then one is logically driven to the conclusion that human efforts at poetry and science and so on are just artefacts of genetic survival with no intrinsic meaning, which is quite wrong. Furthermore, the genes themselves begin to unravel and disappear from view, because they turn out to be products of quantum field theory that have no proper meaning of their own; they are 'explained' by quantum field theory. This also is quite wrong (recall, once again, the lesson of the arch and the lesson of the digital computer: a high-level programming language is not explained by the behaviour of binary logic gates).

Now consider again the statement, 'We are survival machines—robot vehicles blindly programmed to preserve the selfish molecules known as genes.' This statement is sufficiently loosely stated that it can slide all the way from truth to falsehood and back again, depending on what is meant by the opening words, 'we are'.

The statement can serve as a reminder of certain facts about evolutionary biology: 'Human bodies are made of ordinary mindless stuff put together in a configuration that has come about by the Darwinian process.' This is true. In this sense we are, among many other things, survival machines for genes.

On the other hand, the very same 'robot vehicles' statement can be used (and often is used) to say something else:

> If you want to know what a human being really is, then here is the truth of it, the essential point which is prior to all the distractions of the arts and humanities, of Laozi, Euripides, Shakespeare, Goethe, and all ideas dating from before 1859: a human being is a robot vehicle blindly programmed to preserve genes.

This is false. What we *are*, in the sense of what is central about us, is no more captured by a reference to genes than it is captured by a reference to chemical reactions or to entropy. We are indeed physical, and chemical, and biological: well, yes, and so are plenty of other things, but this does not establish what is characteristic of us, or tell us who we are. That is not determined by the mechanism of the exploratory process called evolution, but by the space of possibility that the process explored—especially the aspects of that space that concern social and personal existence. It is this way round because other mechanisms could have arrived at the same result, just as the same high-level computer application can be supported by various different operating systems. If another process had occurred, it would not be able to construct a different meaning to the word 'trust' any more than it could construct a different set of factors of the number twelve. What we are is human: generous and grasping, thoughtful and thoughtless, sensitive and insensitive, forgiving and unforgiving, and so on; and consequently we are appropriate objects of personal consideration, compassion, and justice.

But someone may say, 'that is what I meant: a robot vehicle can be generous or grasping, thoughtful or unthoughtful, and so on.' Fair enough, but in ordinary language that is not what the phrase 'robot vehicle' denotes. The phrase 'robot vehicle to preserve genes', in ordinary language, denotes an entity for which, since no personal qualities have been mentioned, none have been recognized, and therefore one which is not an object of personal consideration, compassion, and justice.

The phrase, 'we are survival machines—robot vehicles blindly programmed to preserve the selfish molecules known as genes' is therefore so slippery as to be thoroughly misleading. By loose use of language it amounts to a misdirection whose uncritical acceptance still fills me with dismay. Though it has been circulating for years, I never seem to hear much willingness to subject it to critical scrutiny.

8
Darwinian Evolution

Having asserted what Darwinian evolution is not, we now consider what it is. The aim is to get a sound overall judgement of what sort of process and sequence is found. We revisit the existence and role of randomness or openness, and look briefly at animal aggression. It is argued that the metaphor of 'eagerness' is better than the metaphor of 'selfishness' when thinking about genes. It is proposed that the journey undergone by life on Earth has not been a mere sequence of events, but a story of notable and genuine increase in richness of expression. Nor has it been merely haphazard, because the very richness it came to express was itself shaped by the patterns that apply at the various levels, including, for example, the level of social existence. The judgement that this is a meaningful story is an intellectually substantial judgement.

I have already discussed Darwinian evolution in the context of the structure of science (Chapter 4), and the difference between purpose and cause (Chapter 7). In this chapter I will present a portrait of the Darwinian process that is intended to convey a sound general impression of what sort of process it is.

All areas of science are a work in progress; zoology and biology are no exception. It is not my purpose in this chapter to look into those areas of the history of life on Earth in which there remain deep scientific unknowns. As I understand it, there are deep puzzles in the origin of life, in the pace of some developments, in the intellectual power of the human brain (which far exceeds what is necessary for survival), and in the presence of such a large amount of developed complexity in general. There are also factors such as genetic drift and epigenetics which modify or complicate the story. I will take the view that better insight into all of these will not overturn the neo-Darwinian picture in general, but one must allow in all areas of science the possibility of the kind of progress that was made in physics between the Newtonian and the quantum accounts. That is, a more complete understanding may change the very terms of discussion, while respecting the accuracy of the earlier insights in their range of applicability.

The aim of the present chapter is to influence contemporary discussion in the following way. I wish to reassert, strongly, the importance of letting the processes of the natural world be what they are, before we rush in with our moral or artistic interpretations. I want to maintain or raise awareness of the difference between logical deduction and artistic impression or moral judgement. I don't think the latter are wrong or out of place, but they are different from the former and should not be mistaken for the former. Having made that clear, I will go on to sketch some general artistic impressions that the story of the natural world suggests to me, and which, I will argue, are a fair reflection of the sort of process that evolution has proved to be.

In the background to this chapter are ongoing arguments that circulate in the popular imagination, one step removed from the demands of academic peer review. Large numbers of people think that they ought not to accept the Darwinian scientific account because they have been told it contradicts theism or the Bible, and equally large numbers think they ought not to accept theism because they have been told it has been rendered incredible by Darwinian evolutionary biology. Both groups are being led astray by leaders who misinterpret what they affirm and lack insight into what they criticize.

And these short-sighted leaders feed off one another. The sad and intellectually ridiculous 'Creation Museum' in Petersburg, Kentucky, USA is a monument to all who have misrepresented the place of Darwinian evolution in the intellectual landscape of theism, and I think it is high time that commentators on all sides stopped doing that. It is simplistic and disingenuous to lay the blame wholly on religious bigotry. When communities oppose the teaching of ordinary evolutionary biology, the cause can often be traced to a form of intellectual goading. People have been presented with the picture of inheritance, variation and natural selection not as a largely correct summary of what goes on in ecosystems, but as an atheist's charter, or as something acutely dangerous to what they hold most dear. And indeed Darwinian ideas were used to justify notions of racial superiority and extermination or chemical castration programmes. No wonder, then, that there was a reaction against it.

A huge boost to anti-evolutionary feeling was produced when, in the nineteenth and early twentieth centuries, people used evolutionary arguments to suggest some classes of people were substandard and ought to be prohibited from breeding, so as to 'improve' the human gene pool. The dismay caused by this sort of talk prompted some non-scientific observers to denounce Darwinian evolution as a godless way of looking at the world, and so it had to be patiently explained, over several decades, that this need not be the case. Meanwhile, another influence came from a form of textual criticism practised in the late nineteenth century, in which, alongside legitimate effort to understand its historical setting, the Bible was repeatedly reinterpreted with too much confidence by scholars who rejected some of its premises. In reaction to this, a section of Christian leadership, especially in America, took recourse to an opposite extreme. They argued that whenever there was the least possibility that a part of the Bible could be read literally, then one was duty-bound so to read it. In fact, of course, such an approach fails to respect the richness of literary practice, and it amounts to an abuse of the Bible, a refusal to allow it to express itself on its own terms.

One would hope that by now such mistakes could have been corrected, and not used to mislead young people about both the Bible and science. But by the time education policy in America had a chance to get to a fair assessment of this, the whole thing was thrown into disarray again when some people saw Darwinian evolution as a grand way to attack theism directly, in intellectual terms, and try to 'blow it out of the water' once and for all. What you see in the work of Richard Dawkins, in this area, is his latching onto an argument due to someone called William Paley writing in the late eighteenth century, someone whom thoughtful modern-day Christians had mostly never heard of,* and whose argument we never used or needed.[2,3] But Dawkins presents this argument of Paley's as if it is the chief cornerstone of faith, and implies that by unpicking its mistakes he can give the appearance of doing serious intellectual work, in which trust in God is revealed to be empty.

Now, to be clear, I don't want to undermine the genuine contribution of either William Paley or Richard Dawkins. Paley was writing from

---

* Not owing to ignorance but to unimportance; Paley earns a mention in a footnote in the history of religion, but that is all.

a pre-Darwinian scientific perspective, and within the constraints of his time he made a positive intellectual contribution. His work was an important resource for the young Darwin; it certainly was a help to him and thus a contribution to science. This is because Paley discussed biology through a wealth of detailed examples and pointed out that, just as the motions of astronomical bodies had their patterns, so also there are many remarkable regularities in the life of plants and animals. Charles Darwin learned much from this and was inspired to find out more.[75] Paley also made a now famous argument by analogy, comparing the natural world to the mechanism of a watch and inferring design, and it is an important progress to know that that argument does not succeed in any direct or simple way.*

Equally, I don't want to underrate the considerable illumination which Richard Dawkins brought to the whole sweep of evolutionary understanding. But his use of the word 'God' is mainly a foil by which he can dismiss various simplistic positions that you can find in folk religion. He does not trouble to mention those parts of the relevant background which might require him to qualify the story he wants to tell—background such as the intellectually serious arguments of the many competent scientists and philosophers who understand evolution perfectly well and do not consider that it contradicts theism.[†] Nor does he ever grapple with the demands of intellectually serious theological reflection. The situation for the latter can be stated easily and is nothing like what Dawkins implies: serious theological reflection *takes the Darwinian process as a datum to be received*. As Keith Ward (Regius Professor of Divinity at the University of Oxford from 1991 to 2004) puts it, 'Most theologians accept that whatever view we take of the physical universe has to be consistent with evolutionary science.'[76, 3]

The failure to place an argument even remotely in its true intellectual context amounts to a failure of scholarship. When it is done in a book

---

* There is a range of opinion on whether or not some more subtle such argument can be made; I suspect not, for reasons given in Chapter 14, and c.f.[40].

† Here are some biologists or palaeontologists who have written about this: Dennis Alexander, Francis Collins, Stephen Jay Gould, David Lack, Ard Louis, Kenneth R. Miller, and Simon Conway Morris. One might also add a significant group of scientists working in other areas, as well as philosophers and theologians who have the necessary grasp of evolutionary biology.

aimed at the general public, it amounts to an exercise in the public misunderstanding of science.

The public discourse on this has become sufficiently misleading that the point I just made needs to be underlined. I don't deny that, within the contemporary Christian community, there is a significant movement, calling itself 'creationist', which fails in its duty to reason, and exists in a state of theological confusion, along with abject denial and refusal to face facts about the genetic and geological record. I already alluded to this. Nevertheless, when I said that 'thoughtful modern-day Christians' never bothered with Paley I was merely stating what is so, since Paley's work, influential in its day, was largely forgotten outside 'creationist' circles, until Dawkins exhumed it. And this forgotten argument was not reinvented or repeated in modern universities, because it only takes a small amount of instruction to get a sound basic grasp of evolutionary biology. One only needs to have a general grasp of developments in science over two centuries since Paley's day; to pay a small amount of attention in school biology lessons. That is why we never bothered with Paley nor were in the least incapable of spotting dodgy arguments.[3] In view of this, I think it is fair to say that, alongside some good science, there is a significant degree of misdirection going on in Dawkins's book *The Blind Watchmaker*.

More recently, so-called intelligent-design arguments have been equally counter-productive, by implying that there is something atheistic about ordinary uncontrolled variation and natural selection, and by announcing 'miracles' without a sufficient effort to interrogate the evidence and construct scientific models in a reasonable way.

In short, this area of science has been presented to large parts of the general public as if the only options were forgivable or disingenuous misperceptions and theism, or great science and atheism. The claim is not that there are no theists who understand the processes of evolution; the claim is that only atheism 'really' understands or 'takes it seriously' and gets the message in full. In the background to this claim are some further philosophical and theological points which I will address in Chapters 10–14.

It is true that certain popular religious ideas are sufficiently out of kilter with what we know about the natural world that they have lost whatever credibility they may once have had. For example, it makes no sense to suggest that the life of animals was once free from pain and death, or that the world was once comfortable and easy until human beings came along and spoiled it.* Also, an idea such as 'Creation' has to be thought about in a nuanced way; the evidence is very strong that the way the ecosystem has come about is not like a bricklayer building a wall nor like a watchmaker constructing a watch. It has more of the feel of a semi-independent creation with a life of its own. This is reminiscent of the way novelists and dramatists experience their creative efforts. Writers commonly report that a character they have imagined takes on a 'life of its own', a certain independent integrity, as if it is not the writer but the imagined character who speaks and decides. I mention this here only to remind the reader that the notion of 'creation' in human experience is itself quite rich, and is very much about conferring an existence on the created item such that the latter stands forth in its own right and acts back on the reader or viewer or listener. The artist begins to stand in a two-way relationship with the very thing he/she has created.

With this in mind, anyone who recognizes that a transcendent creativity gave and gives rise to the universe can also find it perfectly correct and even obvious to also accept the vast amount of evidence for, and glorious insights of, Darwinian evolution by variation and natural selection, and the genetics that accompanies it. I would even say it is my duty to accept such an overwhelmingly well-established, productive, and insightful set of rational deductions, carefully worked out, by God's grace, by my fellows. It is my duty to reason, to rationalism, and to truth, which is part of my duty to God. However, we don't have to accept all the glosses and interpretations that people apply to evolution—we have to carefully choose among them, looking for ones that are insightful and fair to the facts.

The difficulty of Darwinian evolution is that it is said to be random and it appears to be merciless. In fact it is far from random, and its interpretation needs more care.

Randomness is an important factor in evolution, but, as I already commented in Chapter 4, it serves chiefly as a way for small changes to be 'tried out'. The success or otherwise of those small changes (success in enhancing survival and reproduction, that is) is not random but is

---

* The people who wrote the opening book of the Bible were fairly canny about nature and did not suffer from delusions of this kind. They present the world as a *sound* work of art, not an easy one. This basic soundness (Hebrew *tov*) is the reason why science is possible.

determined by the environment. That environment is itself dynamic and chiefly consists in other organisms, and this is why the results have appeared to some investigators to be largely random. In fact, they are not, and this is amply confirmed by widespread *convergence* in evolution—substantially the same 'solutions' to the problems of living (cell chemistry, livers, eyes, etc.) are found independently in more than one place and more than one time.[45] Furthermore, important constraints are set by physical and mathematical facts: this is so important and all-pervasive that it is often taken for granted. The only animals you will find on planet Earth are, obviously, ones that can live there. Ones whose skeletons are strong enough for the local gravity, whose metabolisms are efficient enough for the local oxygen supply, etc. Evolution has only produced, and could only produce, animals like that.*
And such constraints run deep, right into the chemical interaction networks that give rise to cell chemistry,[37] and into the types of protein and DNA that can work because they possess the right combination of stability and flexibility. It is not all determined right down to the finest detail, but by and large evolution does have a sense of direction, and creatures much like mammals, for example, would probably have emerged roughly when they did, even if a meteor had not hastened the extinction of the dinosaurs.

That last statement was deliberately provocative; nobody knows enough to say for sure about an issue like this. The point is simply that the evolutionary sequence is by no means completely haphazard.

The premature or confused statements about purpose that we considered in the previous chapter may be motivated by a proper concern to present the physical process as it is without adornment. When we admire a magnificent tree or a remarkable ant or a wonderful whale, we should not be distracted from the fact that the physical causes leading up to a modern-day tree or ant or whale are the ones set out in standard evolutionary biology. It is a sequence that tends to the result that there be living things with forms and behaviours suited to the proliferation of the DNA which encodes much about those forms and behaviours, and there is no 'essence of magnificence' injected in alongside the physical process. The process itself does the job. But what is really remarkable here is that in fact there is such an essence, at the deepest level, in

---

* This is dynamic and non-linear, of course; the oxygen supply has itself been greatly influenced by plant life.

the very framework of the whole natural order. The 'and that is all' of evolutionary biology takes place within the framework of what is, and of what can be, and this is far from an arbitrary framework.

Evolution is a vast and complex, and really rather wonderful, 'river' flowing into the ecological niches that exist and that become available as more complex organisms come to be.

Thus evolution is not essentially random. Nevertheless, there is randomness in it. The word 'random' seems to carry negative connotations, but it is often the case that what we call 'random' could equally be called 'open'—a point which I already stated in Section 4.3, but it merits repeating here.

It is scientifically and philosophically correct and proper to admit that we never know for sure whether anything is truly random, but I take the view that it is reasonable to say that natural processes are not completely deterministic, and this is what I think. Let me say why.

There are physical events where observations and theoretical models agree that no pattern is either found or to be expected. An example is the set of radioactive atomic nuclei that have decayed after one half-life has elapsed, in any given sample, in ordinary conditions. The set is observed to be distributed randomly across the sample, and the theoretical description (quantum mechanics) proposes that this is indeed so. Faced with such observations, it is true that we cannot tell whether in fact the distribution is fixed by some deterministic pattern, like the decimal digits of the number π. Equally, when large numbers of small changes take place, it is entirely possible that there are correlations among them which one cannot notice by examining them one by one, so that a global pattern results. I think it likely that processes like the latter can happen (there is a modest hint towards that view in the phenomenon of quantum entanglement). However, I also think it fair and reasonable to suppose that there are also events which are open, entirely uncontrolled, and free to go one way or another. The evidence suggests this, and also, I will admit, I am drawn to this view because it helps me to have some notion of why senseless suffering exists in the world.

Randomness comes in whenever there is more than one possible way in which things might develop: the outcome is not dictated. The universe, or one small part of it, had some openness; an opportunity to pick from a range of outcomes; a little freedom. We should celebrate this freedom, I think, because it allowed the exploration that we call

evolution to proceed, and it may be that this openness is exploited by our own neurological processes in order to make our freedom possible.*
However it has a cost: it also leads to genetic defects and to painful illnesses such as cancer.

Now let us consider another aspect: animal aggression. Evolution has too often been misleadingly painted as one big merciless and brutal fight. 'Survival of those best suited to their environment' (which is what 'fittest' means) has been twisted into 'survival of the strongest, the most aggressive'. In fact the story of evolution is much more often one of 'survival of the most adaptable'. That is, perhaps, the main story, in combination with the idea of *ingenious improvisation*. Improvisation is an important theme in evolution, and it is striking that this same concept is also important in human art and in human life.

It is true that weaker, slower gazelles will be more often chased and caught by hungry lions than will stronger, faster gazelles. This is obvious, and such facts are pitiless, but they are not brutal. The lioness does not hate the gazelle but is incapable of empathy for gazelles; the gazelle runs and fights for his life as hard as he can, but does not have any clear idea of a longer life which he might have had if this lioness hadn't got him. What a dying gazelle experiences I do not know, but I guess it is chiefly pain rather than the sort of horror, dismay, and sense of loss which a human can experience. What the lioness does is not murder because an advocate can plead for her the defence of diminished responsibility. Human beings with their greater perception have greater responsibility.

Animal aggression is, on the whole, controlled aggression, and life among the higher animals is, on the whole, social and communal. The whole 'red in tooth and claw' mantra is a distortion; mammal life is largely 'brown in hoof and cud'. This is not to dismiss the bitter struggles that mammals and birds regularly face, against disease, adverse weather, and the trials of migration, for example. They will take advantage of those whose cousinship they cannot recognize, and they will negotiate with, or even nurture, those they do recognize.

That last fact is very much connected with the way variation, inheritance, and natural selection work, and this deserves a further comment.

* This is not the incorrect claim that one can make sense of the notion of responsibly willed choice merely by an appeal to indeterminacy. Rather, the fact that fundamental physics does not present a clockwork model of basic processes, and does present unresolved puzzles of interpretation, leaves the nature of human will an open question.

I have, in the previous chapter, taken issue with the way purpose and cause have been confused in Dawkins's *The Selfish Gene*, but I would like to repeat, as I acknowledged there, that the emphasis on genes does have an important role in understanding the Darwinian process, and I believe Dawkins deserves credit for highlighting this. The essential point is that it is the replicators themselves that are the units on which natural selection acts. There is no built-in friendliness towards whole organisms as such; the genes which tend to proliferate are those whose effect tends to cause more of those same genes to enter the gene pool, no matter how.

I have, in another book, suggested that the phrase 'eager gene' is a better metaphor for this than 'selfish gene'. Let me say why.

In the animated film *Shrek* (PDI/DreamWorks, 2001) there is a memorable scene in which the ogre Shrek asks if anyone knows where to find the ruler Farquaad, and Donkey jumps up and down shouting out 'Pick me! Pick me! Me! Me!'. It is memorable because Shrek does not particularly want to pick Donkey, and because Donkey is brought so much to life by the actor Eddie Murphy, in what must rank as one of the greatest ever animation voice-overs, and because we recognize the scene. It is the scene played out in classrooms, sports fields, magic shows, family parties, toddler groups, wherever a bunch of eager children want to have a go at something that looks fun, though they don't necessarily know what it may involve.

In Richard Dawkins's book the metaphor of 'selfish gene' is adopted in the title and then used extensively throughout, with words such as 'manipulate' and 'tyranny' and so on, as we saw in the previous chapter. I admit that this metaphor has some traction on the truth, and it functioned well as a title. Unfortunately, as I think Dawkins himself has admitted, the metaphor has taken on a life of its own and has been widely misused.

Obviously no one thinks that inanimate molecules such as genes can be either selfish or selfless; the purpose of the phrase was, I judge, to express, in a brief, powerful way, the following idea. The bodily structures, and many of the characteristic behaviours, of individual plants and animals are largely inherited through passing on individual molecules, the DNA, from one generation to another. Self-contained units of DNA, called genes, do specific jobs (cause specific proteins to be constructed in the cell). Because each generation can, and usually does, produce more than one or two offspring, in principle there is the potential for exponential growth in numbers, and therefore many copies of the genes will be generated. During the process, small changes result in many different types of genes being tried out. Now, if you did not think it through carefully, you might imagine that after this has been going on for a long while, through many generations, the genes you will find in any given organism are the ones that promote the health and reproductive success of that very organism. You would be right up to a point, but, in an important sense, you would be wrong, and this is what the phrase 'the selfish gene' is getting at. You would be wrong because the following logic is inexorable: the genes which tend to proliferate are the ones which confer structure or behaviour which results in the proliferation of those same genes. That is all the genes 'care about'. Your genes are not your friends; they are entirely oblivious to you, as a being, and 'care' only about themselves. (Obviously I am lapsing into colourful analogies here, in order to get across the flavour of the idea.) If health and happiness for the host organism result in transmission of that organism's genes to many healthy offspring, then health and happiness is what those genes will tend to confer. Equally, if pain and trouble for the host will, on average, result in behaviour that transmits the genes, then that is what the genes will tend to confer. It is just a question of mathematics and probability.

This use of words like 'host' presents a gene's-eye view of things, and it confers much insight into what is going on in the natural world, including in human behaviour. But it also carries the risk of becoming misleading if it is mistaken for the only view that merits attention. In fact, insofar as genetic proliferation proceeds through high-level processes, *it can only proceed through those high-level processes that respect the patterns of such processes*—patterns such as the combination of sympathy and negotiation, for example, or the combination of disdain and domination. Also, no individual entity in the natural world, whether a piece of DNA, or the ensemble of its copies throughout the biosphere, or a complete living organism, or even a larger group such as a species, can dictate its own future. All are entirely dependent on their wider environment, and this environment includes both physical states of affairs and abstract truths. Furthermore, genetics does not determine behaviour completely; in the case of humans the 'nature verses nurture' question comes out as a roughly 50:50 split in many of the areas we care about.

Coming back now to the phrase 'selfish gene', the problem with it is that it is too loaded, because in human beings selfishness is a moral issue. Another problem is that it seems in practice to bring in the impression that genes are clever, with a machiavellian kind of cleverness, whereas in fact of course they are utterly and completely ignorant of what happens to them. Nor are they immortal, of course: a thing cannot be immortal that was never alive in the first place. A further problem is the assumption, widely made, that the plants and animals expressing these 'selfish genes' must themselves therefore be 'selfish', and that the whole scheme of nature is a huge scheme for the promotion of selfishness. Of course a professional biologist will not make that mistake, and the text of *The Selfish Gene* discusses this very point, namely that the mechanism of the neo-Darwinian process should be expected to lead to certain forms of altruism at the level of whole organisms. Nevertheless, the dynamics of what happens to genes is basically about mathematics and we can pick whatever metaphor best reflects the sense of that mathematics. I would like to offer another phrase which I think does a better job of capturing what the genetic basis of evolution is about.

The phrase is: *the eager gene*.

'Eagerness' is a better term than selfishness partly because it is less loaded; eagerness is not a moral issue in humans. Eagerness is innocent. The phrase remains a metaphor or analogy, of course. Molecules are not really eager. But the fact that genes confer structure and behaviour which simply promotes those same genes, without regard to higher-level issues such as the interests of other organisms, is reminiscent of the eagerness expressed by young children when they are not yet able to process the complex issues that morality involves—the issues that make adults hold back as they consider other people's interests alongside their own. The youngsters say 'Pick me! pick me!' like Donkey; and that is roughly what genes do.

'Eagerness' also better captures the wider picture of what many of us feel is on view in the natural world. What we see is 'Wonderful Life', to quote the phrase adopted by Stephen Jay Gould for the main title of his book on the Burgess Shale and the nature of history (I don't intend by this to agree with the main thesis of that book, but I like the positive title). Plants and animals are, mostly, eager for life, for a chance to be themselves. They do this largely without, it is true, any qualms about the trouble they cause for others, and when they do take trouble for others, it is broadly in proportion to the degree to which they have genes in common. Nevertheless, even after taking this into account, the overall result has not been the promotion of selfishness, and it is wrongly coloured by the unfortunate phrase 'the selfish gene'. What has gone on in biological history, and continues in the world today, is a sort of restless energy, a never-ending trial and error which buds, and buds, and buds again into forms of life that in turn make possible other forms of life, all within the constraints of, and shaped by, mathematics, physics, chemistry, geology, geography, and social science. What has resulted is, in my opinion, wonderful, and I would also call it beautiful.

It is instructive to pause to think about why, in spite of the apparent neutrality of the mechanisms, something beautiful has resulted (not always and everywhere beautiful, but beautiful overall, and worthy of our conservation efforts). It is, as I have explained in previous chapters, owing to the fact that biological forms mirror or embody mathematical and other types of pattern, with their innate symmetries, and the natural world is deeply ingrained with this type of mathematical, physical, and social beauty.

The eager genes, as I have said, carry no built-in friendliness towards organisms as such. Nevertheless, much friendliness has come about because it is woven into the very nature of what can be. The human body is itself an ecosystem and a symbiosis. Our very psyche is a coalition (a learning one, not always at ease with itself). And the exploration that is called evolution has shown a very definite sense of progress towards more and more complex *communities of relating selves*. An ant colony has a large number of ants, but each ant on its own is not much of a 'self'; it is little more than a worker, a cog in the machine. A mammal social group is much richer. It is striking and very interesting that it is a collection of animals, each of which on their own constitutes quite a rich 'self'. It is only in such groups that the concept of *brotherhood* or community is possible. And as such types of living arrive, as they are 'discovered' by the process of evolution (that is exactly the right word: the concept of community was there all along, waiting to be revealed*), it is very notable that far from being pitiless, there are greater and greater signs of concern and long-term efforts to protect those who are struggling.

The lives of individual animals, especially mammals, have plenty of room for fun, while being on the alert for predators, and they have

---

* The logic here is similar to that which we considered in Section 4.2.1 in relation to mathematical reasoning.

periods of calm in between the hard times. Desperate foraging for food in the middle of winter or drought is part of the story but it is far from the whole story. This has been obvious to hunters and farmers since the dawn of human history; no amount of scientific study will make it untrue.

In summary, while professional work in biology, ethology, and the like has continued apace, the marvellous history of the development of life on Earth has been presented in scientific journalism in broad brush with vivid colouring, and the colours are not always well chosen. Outlooks such as the gene's-eye view and adjectives such as 'blind' have proved to be, in important respects, misleading. Obviously an abstract process cannot literally foresee anything, but Dawkins's use of the term 'blind' involves an inference from 'unable to foresee' to 'lacking in direction or purpose', and this is illogical. He asserts that evolution has been an arbitrary, meaningless journey to nowhere, whereas in fact it has been an exploration process marked by increasing depth and signs of meaning. Jellyfish do not care about their offspring; elephants do. Ants are like robots; bonobos are like friends. To repeat yet again the refrain of this book: these facts are not appearances, they are realities. Bonobos really are like friends. And friendship was not invented by natural selection any more than the strength of round columns was. It is a mode of existence that variation was able to discover and natural selection was able to promote. The natural process *discovered* or 'fell upon' friendship and *invented* a way of embodying it. The particular embodiment is contingent, the social relation itself is not. I have argued that this way of seeing things is logically valid. I would like to add that I find it beautiful too.*

In addition to the above, the emphasis on genes sometimes fails to see the wood for the trees. A tiger is a tiger, a creature of innate value and great beauty, not a mere bag for the transport of passenger molecules. And genes do not copy themselves; they require the ecosystem around them, which in turn requires the sun and the sea and the deep patterns of mathematics, physics, chemistry, and (for complex animals), sociology.

---

* The natural process also discovered fear, suspicion, and aggression, which are not beautiful, but equally expressed the fact that such things cannot ultimately dominate, which is.

My point is that the story of life on Earth does not have the overall shape of a merely haphazard journey, and the arguments about symmetry and high-level language open the door to an intellectually valid conclusion that indeed it has not been haphazard. So, the view that evolutionary history is a meaningful story rather than a mere sequence of events is not a fond hope nor a superstitious wish but an intellectually substantial judgement. I don't think one can deduce theism from this, but one can deduce a greater receptivity to some forms of theism.

Now let me speak subjectively and personally for a moment. I *like* being cousins with all the other living things. I am *glad* to be part and parcel of the natural world where I live, a weaving of its strands. I love having my toes in the soil as a well as a head that can voyage among the stars. I think it profoundly beautiful that living things are products of improvisation and tentative exploration and epic perseverance. And I deeply hate and reject any attempt by any one theological or atheological position to try to declare sole ownership of such truths. Natural history is all our history. Scientific knowledge is the property of the entire human race, and of any other race or species for that matter. The attempt to declare a contradiction between science and an intelligently considered religious commitment, as if one were a replacement for the other, is not just illogical but also unjust. It is as illogical as declaring arithmetic to be an alternative to mathematics. It is as unjust as declaring that education is for whites only.

Of course the whole story of the world is also shot through with death. It can have every appearance of an amoral and aimless struggle. But it is, I judge, difficult to sustain the view that that is the final truth of it. I will return to this point in Chapter 17. It seems to me that the story of the evolution of life on Earth has a certain grandeur about it. Darwin was right about that. This vast saga is not something to be ashamed of, but it is something incomplete. It needs a further ingredient, an ingredient of wise and compassionate management and of poetic response. And that is what we are meant to provide.

We look at the world and find it to be pitiless, but it is not, because we are part of it. Natural processes such as earthquakes and continental drift, volcanoes and climate change are just that—natural processes, that is, impersonal. They cannot be either merciless or merciful. If we want to look for signs of mercy in the story of life on Earth, we have to look in *the only place it could possibly be located*—that is, in thinking, mindful, responsive agents. And there it is found—though in a compromised form. So the natural world is not merciless, but *we humans* (and to some extent the other social animals) are the agents in which the possibility of mercy has been embodied. Amongst all these things bright and beautiful, swarming and parasitical, the mercy, and the love, is there, if only we will make it so.

The dinosaurs will not mind having had to die, if we will but sing of them.

9

The Tree

> The life of an ordinary tree is described, in terms of the main physical and chemical processes: carbon capture by photosynthesis; entropy and energy; moisture. The information expressed in the tree comes partly from the DNA and partly from the sunlight. The tree does not push upwards from the ground, but solidifies the air.

The preceding chapters have presented the structure of scientific explanation, and this reflects the structure of the physical world. In subsequent chapters I am going to discuss some issues around the philosophy of knowledge, which is to do with types of knowing and speaking, and some philosophy related to ethics and values. All this will impact on the philosophy of religion, which is the attempt to survey and elucidate what religious discourse (at its best) may and may not mean.

Before we embark on this, the present chapter stands as an interlude in which we will consider and reflect upon something less abstract: a tree. I would like to show you what sort of thing a tree is. This will illustrate more generally what sorts of thing there can be.

When you see a tall tree reaching into the sky, with a thick trunk and strong branches, able to live for centuries, you might think that it grew from a seedling by pasting on layers of cells made from material drawn up from the ground, the way many buildings are made. Tall buildings are often made with cranes sitting on the walls, or with a lift in the middle, used to pull up loads of brick or stone. Similarly, a seedling seems to burst upwards like a fountain in slow motion, the material spraying up out of the soil. It thickens into wood and bark, and it looks as if all the raw material is coming up from the ground, and the tree is growing by extending itself, plastering on layers. It looks as though the tree is pushing its way into an empty space, and its own body, previously built the same way, is providing the support that allows it to reach further and thus fill the space.

Science and Humanity. Andrew Steane, Oxford University Press (2018).
© Andrew Steane. DOI: 10.1093/oso/9780198824589.001.0001

Figure 9.1. Tree in Rosebery Park, Epsom.
Source: Richard Lea.

But in fact it does not happen like that.

In fact the tree slowly occupies a space which is not empty but full of the very air which the tree both needs and uses. The raw material is partly drawn up from below, in the form of water, but a crucial part of the material, the part that allows the tree to form a solid structure, namely carbon, is not drawn up from the ground but plucked out of the air. Almost every atom of carbon in the towering structure that you see, forming about a sixth of the weight of the complete tree, came not from the ground nor from the rain but directly from the air. It did not have to be transported upwards; it was already up there. A tree is way of solidifying part of the air. The leaves are grappling hooks.

The role of carbon in the tree is as *connector and solidifier*. It is the backbone in the molecular skeleton of the tree. Hydrogen can only attach to one thing at a time, but carbon can join things together, because each carbon atom can form multiple chemical bonds. As a result it can be, and is, the connector-atom that joins together everything else in the carbohydrate molecules that make up the solid part of wood. This connector and solidifier is not acquired by progressive pulling up from below, but by simply 'breathing' in from the medium which surrounds the tree and in which the tree lives. It is acquired by direct inspiration.

Carbon dioxide molecules in the surrounding air get attached to molecular structures in the leaves, and various chemical reactions take place there. In a typical reaction, carbon dioxide combines with water to produce a carbohydrate such as sucrose, and oxygen. The oxygen is given off to the atmosphere, and the sucrose is transported in the sap to the growth points of the tree. A healthy tree thus 'breathes' both in and out, acquiring carbon and releasing oxygen.

Furthermore, the information which the tree requires to form a structure (an ordered shape that can grow, rather than an amorphous blob that cannot) is not just in the DNA but also in the sunlight. We often say that the sun provides *energy*, and this is right, up to a point, but almost all of the energy taken in by living things is subsequently given off in the form of low-grade heat. If this were not so, the tree would soon overheat and eventually glow white-hot if it did not first catch fire. What the sun provides is not so much energy as *structured energy*. What plants acquire by absorbing sunlight is *ability to form ordered structures*. This is a more subtle idea than capturing carbon, but equally telling and thought-provoking.

All the basic chemical reactions described above require energy. An energy 'kick' is provided when photons (parcels of light) from the sun are absorbed in the leaf. Thus the tree exists and grows, and can only come into existence and grow, in the context of an intimate interaction with something quite unlike a tree, something that endlessly sends out its energy and its right-to-grow vouchers.

The idea of 'right-to-grow' vouchers is my attempt to convey the basic physics of *entropy* that is involved here. *Entropy* is the word we use to describe the inexorable tendency of structures to break, and of energy to get dispersed and thus degraded into forms that are not useful. Without something to counter this tendency, everything slowly turns into either a dusty desert or a diffuse gas.

The only way that structures such as the bodies of plants and animals can form is if they pay an entropy cost: they have to cause something else nearby to increase in entropy. In more everyday language, the way that ordered structures are created in the universe is always by a process that *removes disorder*, and disorder cannot be destroyed; it is only ever increased or simply transported from one place to another. So if you want more order in a given place, you have to arrange for disorder to be transported away from that place.

This is one of the fundamental physical laws of the universe. It is called the second law of thermodynamics. No physical entity, on its own, isolated from other things, can ever manage to increase the degree to which it is structured (and thus decrease its own entropy). Structure can only be acquired by doing a deal with other bodies or parts of the universe: something has to play the role of garbage truck, carrying the entropy away.

The way plants do this is by 'doing a deal' with the sunlight and the air. The central fact of the thermal physics of life on Earth is that the sun is considerably hotter than the Earth. This means that plants can exploit the following 'trick'. They take in high-grade energy from the sunlight, and give off low-grade energy to the surrounding air. The energy absorbed from the visible light from the sun comes with small amounts of 'garbage' or disorder; the energy subsequently given off to the air carries away larger amounts of 'garbage' or disorder. Thus plants arrange the 'garbage disposal' that is at the heart of their ability to bring structure where previously there was none.* Now let's stand back and look at the whole process again: the life of a tree.

---

* A science text can make this argument more precise by talking about *temperature*. Temperature is defined as the measure of how much energy is supplied for a given amount of garbage. The hotter a body is, the better quality is the energy it provides. But we have already exceeded the level of detail that is needed to get the main point here.

The Tree

Before a tree grows there is randomness, or absence of pattern, in the amorphous water and air. This randomness is extracted by transferring it onto heat energy 'banknotes' which are then given out to the planet and eventually radiated into space. They leave behind a pattern: cells and twigs and leaves and branches. The DNA gives a specific set of instructions, and the sunlight gives a generic 'permission to make something', which enables the instructions to be followed and writ large in the shape of the tree.

The information that is central to there being a tree is not all encoded in the DNA. The DNA simply gives instructions for a plant of this type to be built, as long as the raw ingredients are there. But in order to amplify the DNA, in order for something large and tree-like to form, there must be both the carbon in the air and also the structured energy contributions from the sunlight—the 'right-to-grow' vouchers that we call photons. The carbon is all up there already, waiting to be grappled with. The tree does not move upwards into an empty space, but rather solidifies specific parts of a region that is already full.

Now take a long look at an actual tree known to you, one visible through your window, or one you can go and find and stare at. Your favourite tree. Fresh layers of cells are being generated at the growing points. The instructions for cell division and growth are encoded in the DNA, and the tree is continually drawing up raw materials from below. The opportunity to follow the instructions is given by a steady supply of 'right-to-grow' vouchers from the sun, and the tree extends itself by catching hold of carbon from the air. The carbon atoms are the connectors that join together everything that forms the body of the tree.

Sky Hooks

I used to think they built themselves up from the bottom,
the little shoot drawing up nutrients and spilling upward
into layers of cells, like a slender green volcano.
You would think they did it like that—
construction in careful stages, built up from
bread alone, like a proof; the existing
structure a conveyor belt for pulling up
simple stuff from below, till the edifice
can tower above its origins. That is the sound
of cultured voices in the current air.

But the thrashing castles do not do it like that.

They lay hold of the sky. They appear out of air,
out of thin air.
Deep in the leaf a chloroplast organelle
hooks a carbon dioxide from the floating
clear supply and, with a kick provided by
gratuitous beaming of an incandescent ball
off in outer space, it cracks off a carbon and
turns water into sweet wine ($C_6H_{12}O_6$)
and oxygen. The branches breathe themselves into being,
hardening in the air like ice fronding the window.

The drawn-up fluid is required, and the structure
must obey all the rules of structures, but
the roots grow because the leaves go fishing in the sky.
It is a matter of catalysed inspiration, seventy per cent air
and thirty per cent water. Of all this, half is breathed back out,
and the final massive form—cottonwood, beech,
giant sequoia, yellow meranti—is five parts gathered
rain drawn up through its own being, and
one part Connector plucked out of the element
where it lives and moves and has its being.

PART II
VALUE AND MEANING

10
What Science Can and Cannot Do

> First the history of science is sketched. Then science's limits are discussed, especially issues of completeness and cogency. In what sense is science incomplete? How can other ways of speaking be cogent? These issues are to do with metaphysics, value, and meaning. The meaning of a text does not lie in the letters. The abstract analysis of a friend does not make contact with their friendship. I offer opening gestures towards what religious language can mean.

This book has the subtitle 'A Humane Philosophy of Science and Religion' in order to signal that it contains philosophical reflection, and in order to suggest the notion of a humane outlook which respects science.

I began with the structure of science itself, and the fact that one can respect the insight on show in all scientific disciplines and also in the arts and humanities. With this in mind, it is correct and appropriate to bring to bear all such insights when forming a judgement about the role that things play in the grand story of the physical world, and hence what may be their purpose. However, forming views about purpose involves bringing in the consideration of *value*, or what is valuable, and this is not determined by the analysis of the microscopic processes that have caused things to come to be as they are.

In the next chapter I will discuss the notion of value at greater length, and this will inform subsequent chapters. Part of the aim will be to show the sense in which reason, that great champion of science, cannot on its own determine what values are most important. The present chapter will open up this same territory starting from the scientific 'end', so to speak. I want to continue to be positive, serious, and thorough about the wide range of what science can do, while also getting into view the types of thing that science cannot do but which are worth doing.

I seek to show that the sort of passive analysis that science engages in is not the form of engagement with truth that is appropriate in all circumstances. Scientific analysis is fruitful and important, but it is not the whole of what is needed in an outlook which tries to do justice to the whole of reality and to make an honourable contribution to thought, education, political process, technology, art, and all the other ways in which we can try to promote a better future.

Thus, the present chapter will focus on science and consider the sense in which it contributes part but not the whole of what merits our attention. The widening-out which is involved when we look beyond science is very much about discerning what is valuable and what is not; what merits our attention and action and what does not. It is also about receiving our aesthetic experience and allowing it to mould us appropriately; also about grappling with our sense of justice and injustice. All these are to do with the notion of *meaning*. We need a word which we can agree signifies the attempt to get meaning right, and live accordingly. For a long time, the word 'religion' was that word, but historic and cultural currents have coloured the connotations of 'religion' so heavily that it is now all but lost to us under the baggage of religiosity. Nevertheless I included it in the subtitle of his book because I think it is better to try to rescue the word, rather than abandon it.* I mention this here because the present chapter will offer some hints and pointers towards what healthy religion is like. In particular, I will make initial gestures towards the idea that the transcendent is not only a sort of aesthetic pressure, but also that which impinges on us as people in personally significant ways. In short, we are not merely supported by a glorious mystery, but also known and called

---

* This footnote is addressed to the modern practice of distancing oneself from the word 'religion' in order to say that we don't need religion in order to find meaning. The problem with this practice is that it assumes some unstated definition of the word 'religion' which usually amounts to little more than prejudice. If 'religion' is *by definition* something other than wise, intelligent, and careful about rationality, then really the word has become a form of name-calling. But if by 'religion' we mean that form of spirituality that is grown-up enough to take on serious intellectual work, and other work, to question its own assumptions and to put into practice its better conclusions, then we have a fair use of language. The demonization of 'religion' is suspect because such demonization is, all too often, a ploy whereby one assessment of meaning gets to assert itself as the prior or natural or default or only objective assessment, or the one that 'reasonable' people must find. This encourages people to indulge in ways of thinking that judge others not by how they behave but by what one suspects are their motives. This is a serious issue, one which leads to much injustice. However, a lengthy analysis would be required to present this properly. This footnote merely signals that such an analysis would, I think, do us all a service.

upon. The willingness to affirm this I shall call *theism*. Its denial I shall call *atheism*. Note that theism does not, in my usage, imply that one thinks there exists a supernatural powerful entity outside the universe and overseeing it. That, I think, would be a form of idolatry. Rather, theism simply affirms the appropriateness of an attitude of trust towards the real foundation and enarching shaper of all things, without trying to capture that One in any simple picture. To recognize God is to adopt an attitude of trust and of asking, seeking, knocking on the door. In Christian experience it is also the experience of coming to be known. By 'shaper' here I have in mind an analogy with the shaping influence of symmetry principles in fundamental physics, but of course this is rather a loose analogy. It is this analogy pursued into the dynamics of personal interactions, where the terms of discourse, and the appropriate behaviours, are very different from hypothesis-forming and testing. Please read on for whatever greater clarity may emerge!

By now the reader is alert to the fact that I am not willing to dismiss theism quickly, and in fact I think that an intelligent and generous theism can be both morally attractive and intellectually respectable. However, I am trying to keep this view on a 'low burner' rather than promote it too vociferously; I want to give the reader room to think, room to explore, room to breathe. That idea of *opening out*, or finding permission to realize one's whole self, is close to my whole aim, in fact.

In Chapter 14 I will address a point which is at the heart of what religious language does and does not mean. This concerns the view or intuition, often felt quite strongly by people with a good aptitude for science, that theistic religious language cannot possibly be anything other than superfluous and misconceived, because it does not contribute a simplifying concept that we can bring in and put to work in a scientific treatise. The present chapter will introduce some ideas which will help the reader to follow my reply that is presented in Chapter 14.

One thing that troubles me, and that bothers sceptics, is that often there is, associated with religion and beliefs, a style of thinking—a mindset—that is at odds with science and rationality. The main part of the answer to this is that religious thinking does not need to be irrational and credulous. Who says it can't it be wise and well-constructed? Of course it can! And it can have the fresh air of basic honesty, and the tang of zest for life, too. It is not my purpose to discuss issues of science and religion in general in this chapter, but I will survey some central items, with a view to allowing both atheistic and theistic voices a chance to express what they care about.

The next section sets the scene with a consideration of the history of science. I offer general observations of well-known facts, and also some remarks on aspects that are often ignored or overlooked or misunderstood. For example, the political and civil arrangements in any given time and place are a significant factor in what kind of intellectual work can be done, and by whom.

The main portion of the chapter (Section 10.2) is concerned with the nature of what science achieves; I want especially to engage with the widely felt impression that scientific reasoning will eventually lead to a correct and complete account of everything that happens, in such a way that there will be no role for any transcendent reality. According to this idea, science is already largely displacing religious ways of thinking and eventually this displacement will be complete; theistic religion will not be able to add anything cogent or helpful for people who care about clear thinking, except possibly some interesting works of art. As I say, I want to engage with this impression, to recognize and understand it, but with a view to showing why it is a false impression. My answer will not be to argue that scientific models have gaps, but rather to say that they themselves are situated in a larger framework. This framework is as loosely indicated in Figure 3.2. In earlier chapters we looked at how this framework operates within science; in the present chapter we explore the limits of science, of what it can do. This is related to the discussion of values to be presented in the subsequent chapter. By limits of 'science' here I mean the limits of the competence of abstract model-making to answer legitimate questions about value and how we should live. This is loosely analogous to the difference between asking, of a piece of music, 'does it satisfy rules of strict counterpoint?', and asking, 'does it succeed in aesthetic terms?'. Of an ordinary physical happening of any kind, one may ask not only, 'what are the processes going on?', but also, 'what does it amount to? what does it say?'. The aim in this chapter is not to answer such a question, but to point out that it may be asked.

10.1 A Brief Historical Survey

Let us look briefly at the historical development of science. On one hand, the attempt to observe the natural world carefully and to respect its patterns goes right back into prehistory, but on the other, that interest was not organized into what we now call 'science' until relatively recently—to a limited extent in ancient Greece and Rome, and more completely in the late Middle Ages in Europe and the Middle East, and growing from there.

The contribution of people such as Plato (c.400 BCE) and Aristotle (c.350 BCE) is usually called 'philosophy' (the love of wisdom) and this is a correct term, but this does not mean it is unconnected to religious commitment. The Greek schools that were most scientifically productive were, broadly speaking, those which saw the world as the product of a rational organizing principle or *arche*, and which recommended, like Aristotle, a sense of respect towards God expressed in non-mythological, monotheistic terms. Aristotle himself might be 'claimed' into either an atheist or a theist camp, since the qualities he attributes to God might be interpreted by a modern reader in either direction. Aristotle wrote of God as one who could eternally contemplate and reason, but not as one who would be concerned with the lives of individual people. One should, of course, allow Aristotle to be himself rather than commandeer him in support of later positions or questions that are one step removed from his expressed thought. It is fair to say that the state of contemplation he attributes to God is one he understands to be a positive condition to which he himself was drawn. Aristotle was careful to avoid dubious mythological religious claims and the ill effect of being too bothered by them, and in this he was either a good atheist or a good theist. One can respect the careful way he carried out his observations of the natural world without being distracted into superstitious ideas about cause and effect.

The flowering of natural philosophy in ancient Greece and later in Hellenistic Egypt included the work of Hippocrates and Euclid and many others. In India and China also there were notable mathematical, astronomical, and technological achievements. Nevertheless, in the period of roughly four hundred and fifty years between Aristotle and Claudius Ptolemy (c.120 CE) the errors in many of Aristotle's basic ideas about motion were not corrected. This fact helps us to calibrate the rate of progress. New knowledge was hard to achieve and often required conditions of political stability in order that people could learn and build on previous work.

In the first few centuries CE, Christian influence grew in Europe in conditions of travail as groups struggled for the right to existence in the Roman state. This period is now assessed in wildly different terms by different modern writers. Some present Christianity as having essentially a pernicious effect on scientific progress. This is not true, since Christianity improved the quality of life of its adherents* and in the long term this helped progress in science too. Overall, the long-drawn-out ending of the Roman Empire was a complex period, not easy to analyse. My own limited understanding is that Christian reflection gave a new impetus to champion the rights of the poor, and this led to a focus on issues of governance and civil law. The very idea that the civic authorities were themselves sacred, which lay at the heart of the Roman world view, had to be overcome. These issues drew the efforts of many of the thinkers of the age. There was also all the human folly that attends all aspects of human life, including religion, and philosophy is not immune.

The Western philosophical tradition continued in places like Alexandria. Christian and pagan philosophers met together and sometimes collaborated, sometimes attacked one another's ideas. Both these groups had religious commitments and the result was what has become familiar: some ceased to respect others, but the best and most thoughtful of both parties could respect the other while continuing to critique ideas as objectively as they could. In particular, the Christian philosopher John Philoponus (c.500 CE) learned from Aristotle's work, critiqued and in some areas significantly exceeded it.[75]

In the sixth and seventh centuries, clergy working on civil law in Europe initiated a large-scale programme in which the physical world was stripped of intentionality—'leaving it to be just that, a physical world', as Sidentop puts it.[63] They also began to replace criminal verdicts based on physical combat or oaths with a search for evidence. Under Charlemagne the clegy instituted wide-ranging legal reforms which gave a legal framework around the idea of people as free individuals in their own right. In this they far exceeded the ancient Greeks, for whom barely one person per family got that sort of respect and legal recognition. All these reforms laid the groundwork for the growth of a more universal form of science, in which it could become the practice of a large number of people rather than a few philosophers and their schools.

The period from approximately the eighth to the thirteenth century, known as the Islamic Golden Age, saw a flourishing of science and

---

* When, that is, they were not being murdered *en masse* at the order of state officials.

mathematics in the Islamic world. This was helped by a fine example of businesslike scholarship in the work of Syriac-speaking Christians who translated works of ancient philosophy from Greek to Arabic. Hunayn ibn Ishaq (809–73) has been described by the historian Withington as 'the Erasmus of the Arabic Renaissance.'[13] Muslim scholars such as Ibn al-Haytham (Alhazen) (c.965–1040) and Jābir ibn Hayyān either pioneered or greatly extended the empirical method. In Europe Christian scholars such as Robert Grosseteste (c.1168–1253) and Roger Bacon (c.1220–1292) energetically and insightfully promoted areas such as optics and astronomy, and further emphasized empirical methods.

A recent study of Grosseteste's work has shown that at the end of the twelfth century he was grappling with the idea of a single physical theory that treated astronomical bodies and affairs on Earth in a single framework.[43] He brought some modest mathematical analysis to bear on this, and used it to support a cosmological model consisting of an explosion of a primordial sort of light which rarefies and gives rise to the various parts of the cosmos as it was then conceived, in a process reminiscent of a phase transition. This is high-quality scientific thinking. Grosseteste also had sophisticated ideas about the nature of matter, reasoning that infinitesimal particles could not on their own account for the volume and stability of matter, since they have no extension—something continuous like light would be needed. His statement, 'Whatever sustains extension in matter is either light or participates in some part in the properties of light', is penetrating and correct.

Of equal, or perhaps greater, importance than individual practitioners* was the whole infrastructure of the medieval universities, which made the growth of what we might now call a scientific community possible. Such organized attempts to understand natural phenomena were not regarded as an alternative to religion, but rather as an expression of one of the intellectual virtues which good religion commends and embraces.[33, 75]

---

* If only for the sake of combating Internet myths about the 'Dark' ages, here is a list of notable others investigating the natural world with intelligence and care in the medieval period: Albertus Magnus, Robert Grosseteste, Roger Bacon, William of Ockham, John Peckham, Duns Scotus, Thomas Bradwardine, Walter Burley, William Heytesbury, Richard Swineshead, John Dumbleton, Richard of Wallingford, Nicholas Oresme, Jean Buridan, Nicholas of Cusa.

From approximately the thirteenth to the seventeenth centuries, a major component of the efforts in universities in Europe was the learning method that came to be called *scholasticism*. This refers not to any particular idea, but rather to a method of testing or extending knowledge based on dialectical reasoning. In this period, logical skills were honed, which fed into the scientific progress in later periods, but there was also too great a trust placed in Aristotle's genius, and like all academic pursuits the scholastic method had its characteristic ways of going wrong: it could descend into endless nitpicking over existing ideas.

Eventually the scientific ideas of ancient Greece and Rome (Aristotle, Ptolemy), alongside further progress from Christian thinkers (Philoponus, Grosseteste, Roger Bacon, etc.) and the mathematical progress in the Islamic world were absorbed by the wider Christian scholarly community, which in turn further extended the empirical scientific method at the end of the sixteenth century. People such as Copernicus and Galileo, Kepler and Francis Bacon were serious about and fully committed to their Christian theistic convictions; they were not merely following social conventions of their time (did any great scientist ever do that?). In particular, Galileo Galilei (1564–1642) was serious about his theism, as is clear from his private letters and his insightful comments on the proper interpretation of the Bible. Since a potent myth has grown up around what happened to him, it is worth briefly reviewing his case.

Galileo was a lively and pioneering scientist, who made central contributions to several areas, especially mechanics and astronomy, and who developed and demonstrated the importance of controlled experimentation. He was also argumentative and slow to acknowledge his scholarly sources, and he tended to mock people and overstate his case.[34, 33] In his late sixties he was brought to trial and found guilty of promoting 'error' by a church court, and sentenced to remain at his villa and surrounding land for the rest of his life—a harsh punishment, considering he had done no harm except insult some people. The church establishment of the day got it wrong in his case, but this was an isolated incident that took place about four hundred years ago, Galileo continued his scientific work and was encouraged to do so by leading churchmen, and most of Christian culture has been supportive to science (from its promotion in the medieval curriculum to the extensive establishment of universities throughout Europe, and through the financial support given to individual practitioners). So why is this episode so often brought up? It is because it has become a representative case of what happens when free-ranging enquiry meets self-interest and controlling thinking. It was so invoked in a memorable play written in 1939 by Bertolt Brecht, and this partly explains the ongoing fascination with the 'Galileo affair'.

The Galileo trial and what led up to it is a truly fascinating mix of human character, wisdom and folly, concern for truth, bad and good behaviour, and the history of ideas. For all these reasons it has been repeatedly examined by historians and also by people with political or other agendas (such as Brecht). For example, it is often characterized as a clash in which science and religion came into conflict, but this characterization is not really sustainable, because it does not pay attention to the breadth of what religion is, including its good as well as its bad qualities. Galileo himself should be regarded as both a religious and a scientific person, because that is what he was. Christian commitment was not, for Galileo, a matter of paying mere lip service to religious convention. He clashed with the leadership of the Roman* church because he wanted to reform that church, not overturn it.[64, 34] In short, this was an internal Christian dispute.

Galileo was skilled and passionate about science, and also interested in helping the Roman Catholic church both to continue its support for science and to treat the Bible correctly, which includes, for example, not getting hung up on literal readings of figures of speech. Meanwhile, among the Cardinals and other leaders of the Catholic church there were some who were interested in and very well informed about astronomy. One should not buy the notion that one person is 'more Christian' than another merely on the basis of their official role or day job; in this debate Galileo is as Christian as anyone else.

One should also not sit in twenty-first-century judgement on seventeenth-century officials without first taking the trouble to understand what concerns they had. Large among those concerns was their awareness of the turbulent history of Europe, in which armed mobs and invading armies had repeatedly swept away both the lives of ordinary people and many laboriously constructed works of natural philosophy,

---

* I say 'Roman' here because these events took place after the Protestant Reformation had been ongoing for a century; obedience to Roman Catholic teaching was not such a central issue for Protestant countries or groups.

art and architecture, and carefully accumulated knowledge. Precarious situations of that kind are difficult to manage. The result was that civic and religious leaders saw their duty very much in terms of preservation, so that they required high standards of proof before a new idea about the natural world would be granted the right to free expression. Their attitude was not wholly unlike the modern-day practice of establishing a national curriculum for schools. The restraint required of Galileo was not on what he might write about and study but on what he might teach as an established fact.

Some of Galileo's opponents were unprincipled and some were intellectually lazy, but this is not enough on its own to explain the outcome. The outcome (i.e. the long struggle, trial, and conviction for Galileo) went the way it did partly because the evidence about the motion of the Earth was not yet as compelling as Galileo felt it to be. The data collected by Galileo and Kepler was consistent with Tycho Brahe's model, as well as Copernicus's heliocentric model, while the lack of observable parallax of the stars tended to lend support to the former (Brahe).* Therefore, at the time, it was perfectly accurate to say that the scientific case was not settled concerning the motion of the Earth. Other issues, such as the causes of the tides, were also relevant and not yet understood by anyone.

Our duty to truth always involves a balancing act between preserving knowledge and allowing mistakes to be corrected and new discoveries to be made. It is simply incorrect to regard one side of that balance as 'religious' and the other 'scientific'; in fact both parts are religious and both are scientific. The Roman church leadership did need to be shaken out of a too inward-looking and conservative way of thinking, but one should also give it credit for its efforts in support of science. Church leaders maintained a steady commitment to furthering scholarship in general, and invested large amounts of resources in (what we now call) science. Giovanni Battista Riccioli (1598–1671) and others carefully repeated and added to Galileo's work; Jesuit astronomers enthusiastically adopted and taught the use of the telescope; universities continued

---

* *Parallax* is the phenomenon in which the direction to a given object, from a given viewpoint, changes when the viewer moves. This was understood perfectly well by astronomers in the seventeenth century.

to grow. Galileo's case is like a rare stumble in a ballet of long duration and many performers.

Returning now to the wider picture, the English polymath Francis Bacon (1561–1626) is a central figure in bringing European thinking out of a jumble of natural magic, scholastic debate, and humanism. Bacon brought clarity to two large issues.

In Christian Europe there had long been anxiety about allowing free reign to curiosity about natural phenomena, because such curiosity has no limit, and thus it risks becoming intemperate, or mindless, like a tourist photographing everything and seeing nothing.* In this way intemperate curiosity can displace such goods as an all-round genuine interest in and concern for others—the attitude which was called *charity*, or in modern English, love. Writing within, and formed by, this world view, Bacon helped all, both within and outside the Church, to see the pure pursuit of knowledge as a good as long as it was 'for the benefit and use of life'. He then showed how the empirical interest of the alchemists and the pure logical interest of the scholastics could be brought together, and thus produce what we now call the scientific method. Francis Bacon did not originate the scientific method—his earlier namesake, and others, had already championed it—but he largely settled the remaining hesitations.

If one traces the course of science, in its various disciplines, one finds an interesting and diverse story in which people committed to faith in God have played a strong part, alongside other people. It is nothing like the mythical version which has seeped into popular culture, in which clergy continually act like King Canute, feebly and ignobly opposing the inevitable march of freedom and knowledge (I mean the mythical Canute that tried to repel the waves, not the historical one who knew he couldn't do that). Atheists are right to insist on the value of free enquiry, but they can be fairly accused of encouraging this misrepresentation of the history of science in the popular mind. Consider, for example, the form of writing in which a group called 'scientists' is contrasted with another group called 'the faithful' or 'the church', thus implicitly denying or ignoring the fact that these strands of human activity interact in an interesting and complex way, and overlap considerably. In fact,

---

* A related modern disease has led to science journals becoming clogged up with a hotch-potch of research results having little lasting insight or value.

faithful people, many of them clergy, were often the ones looking for and publishing the scientific knowledge. Indeed, the established church in England had a near-monopoly of English science in the nineteenth century. Its failure was not in doing good science, but in being slow to drop such a monopoly. It was this social issue, more than the content of scientific ideas, that was driving the efforts of people like T. H. Huxley to oppose the religious establishment—see, for example, the careful and thorough study by J. V. Jensen.[36] Huxley was fighting for a system which would allow people to be paid to do science simply on the basis of whether they were good at it, not on some other basis such as patronage.

In short, the difficult relationship between science and the religious establishment at the turn of the twentieth century was much more like the insistence of a grown-up child on his or her independence than it was like an escape from prison.

More generally, the relationship between intellectually serious religious conviction and science is indicated by a broader set of people. These are not primarily those clergy who were also scientists but, more widely, all those natural philosophers and scientists who regarded their life in religious terms and considered science to be a part of it. Care is needed to assess this, because the impact of such conviction is not indicated only by the people who are most vocal about it, but also by those who show evidence of having lived by it in a deeper sense, and their verbal testimony is usually more carefully and hence quietly expressed. A list of people who fit that description includes a notable number of really impressive scientists—I mean either Nobel-winning or Nobel prize-worthy. This is not to prove anything by weight of numbers; my point is merely that there are forms of religious conviction which show great integrity and which do not at all compromise a scientific contribution.

There is also a misguided and irrational fringe in most religious communities, a fringe that can be very vocal. This includes, for example, the idiotic and unprincipled activity called 'creation science'. That activity stands, in relation to valid religious thought, something like the way pseudo-science stands in relation to valid science.

The twentieth century brought an explosion of scientific enquiry, and, perfectly properly, that explosion carried science outside the province of churches and other religious establishments where it began. This is not because science escaped from something that opposed it, but because it was handed on by a community that nurtured it.

10.2 Completeness and Cogency

Next I would like to acknowledge what is perhaps the single greatest difficulty for, or objection felt by, a scientific thinker when coming up against theistic language. This is the objection that science can in principle, and will eventually, offer an adequate explanation for absolutely anything we can possibly think or encounter, and this explanation will not need to invoke a supernatural being. This is the point I will address more fully in Chapter 14. Here I will make some observations on what science can and cannot do.

First, the kind of theism that can command intellectual respect is not so much about 'a supernatural being' as about being itself. As Rowan Williams puts it in a commentary on Mark's gospel, '[God] works outwards from the heart of being into the life of every day—not inwards from some distant heaven.'[81] I will return to this point in the rest of the book. It is centrally important. Atheism is correct to resist the notion of a powerful authority figure overseeing the universe at one step removed; such a figure does not command respect. So good religion does not buy into that sort of picture either. But it is a picture which is often lurking in people's imagination whenever the word 'God' is used, so unless we are careful to explain what we mean when recognizing God, we are liable to be misunderstood.

Next, the position is not that science explains some things but not everything. Rather, science, on its own, does not explain anything at all. It does not explain; but it is part of a larger explanation. Science teaches us about connections between one thing and another, and it describes, but its chain of connections eventually runs out, at both ends—I mean the ends of *depth* and *meaning*. These are often called *origins* and *purpose*. I use the word 'depth' in order get at the fact that when we talk about the origins of the universe or of anything in the universe, we don't just mean beginnings a long time ago. By origins we mean roots, and this includes whatever it is that somehow enables there to be physical existence right now in the present moment. This is a metaphysical question, and I admit I do not know how to raise it in such a way that I am confident that I am asking a well-posed question. But this difficulty merely underlines the sense of mystery, the sense of *knowing that one does not know*.

The equations of physics are beautiful and convey much insight, but we do not know how the great gulf between mathematical possibility and physical realization is crossed, nor do we have even the remotest idea how it could be crossed. And yet it *is* crossed, day in, day out, as all things continue to be what they are. Of course, one could just take the fact of existence for granted, but there is nothing improper or unworthy about enjoying a sense of marvel at it.[42] This enjoyment is, from a scientific point of view, perfectly appropriate. We may describe the 'fabric of space-time' using something like quantum field theory or string theory, but we have no idea, from science alone, what these quantum fields or strings or whatever really are, nor do we know why they should come to exist in the first place or persist even for a microsecond.

Thus all physical things have a present existence that is, at root, mysterious, and inaccessible to science or the scientific method. Scientific 'explanations' are lines of connection starting out from this mystery and then invoking the assumption that the universe is somehow shot through with deep pattern.

Furthermore, we cannot say why reason is better than unreason, using any argument from physical objects such as molecules or neurons. We can only decide to champion reason and rationality. Any argument along the lines that reason is self-consistent merely invites the reply (from some lunatic who wants to champion unreason), 'I don't want to have consistency'. We can only *choose* to prefer reason, by an act of the will; nothing can logically require it. We just passionately affirm that reason is best, and by this act of faith we make science possible.*

Moreover, the assumption that the universe is reasonable is itself a much richer assumption than many people realize. This was the subject of Chapters 3 to 6. The general idea of formulating a bottom-up description of the universe is, it turns out, too vague on its own; it does not constrain scientific ideas enough to allow us to make progress. We need higher-level guiding principles such as symmetry, and these appear throughout science. It is much too early for us to make the connection in full, but it is conceivable that other, richer guiding principles can also be invoked in order to make scientific sense of the universe—the principle of love, for example, which is the principle that the universe be one in which love can be embodied and lived by. (And since love is voluntary, such a universe also has the possibility of hate.)

---

* This is a genuine choice. People can also, and often do, choose unreason, when they buy the claims of homeopathy, for example, or when they believe market forces to be an unqualified good, or when they prefer eloquence over honesty, or when they are receptive to the type of simplicity that prejudice offers, etc.

The previous paragraph addresses the fact that analysis can in principle meet or engage with statements about large-scale purpose, but it cannot provide those statements. We all suspect that there is some purpose to our lives—except for those unhappy people who are, I hope, receiving comfort and help to recover from deep depression. I don't necessarily mean some grand purpose; I just mean that which motivates simple daily efforts to achieve something beyond mere continuation. We very definitely think we each have an 'in order that', not just a 'because of'. And many of us feel that attempts to describe this 'in order that' using finite language are barren. For example, the purpose of my life can't be simply 'in order that' I beget and raise children, because then their purpose is the same, and the whole thing turns out to be meaningless. My and their purpose must be something more metaphysical. It must be to leave the world better than we found it, or to declare some truth—whether by speaking it or by living it. Science is part of the truth, but it cannot tell us what is our truth, the one we are meant to embody. Is my purpose to find out all knowledge? Is it to speak exquisite poetry, to sing in tongues of liquid gold? Is it to proclaim political reform? Is to be trusting, to embody trust? Is to give away all my possessions, even my very body? Well, someone has said that all those worthy things are in fact worthless without love. So it is not so easy to say what our purpose is. It seems to take us out of ourselves, out of the whole merry world of mixed-up humanity, way up into the stars and beyond them. Only once we have planted our standard firmly at infinity can we can get on with ordinary practical living in the here and now.

So the reach of science runs out at both ends, and for the sake of clarity I must repeat it: *science does not, on its own, fully explain anything whatsoever.* Its wonderful descriptions and insights are part of a larger explanation.

I fully acknowledge that, notwithstanding the above, one can still imagine a complete description of the world that has at its roots atheist philosophy rather than theist reaching for God. Such a description would be based on one type of faith, and it can be complete: it can give an account of everything we ever experience. However, that is no guarantee of its correctness. A description can be complete in its own terms, and yet fail to include a central insight. For example, a description of what you see on a cinema screen could be offered completely in two-dimensional language. The action in every frame could be described in terms of planar geometry. It could be so described by someone who had never become aware of three dimensions, for example. Such a description is complete, in its own terms, and yet it is missing a crucial insight: the insight of three dimensions. What is on show is a projection of a three-dimensional scene, and as soon as this is grasped, the whole thing makes much more sense. It would come like a flash of marvellous insight.

Once you grasp this point, it is easy to find further examples. The completeness of the atheist description is not, in itself, a guarantor of its correctness, and nor is the completeness of any other description. No one can announce that they have at last swept away all the nonsense and discovered the truth of things. Living with doubt and the desire to learn is the only valid form of life.

Here is another illustration about complete descriptions. One could imagine a detailed analysis of the markings on the Rosetta Stone that had every appearance of completeness—it left nothing out and was as thorough as it is possible to be—and yet which failed to mention that the markings were symbolic indicators—words—in a language that describes other things. The markings on the stone in fact describe things that are nothing like bits of stone—things like kings and priests, temples, corn, and feasts. Of course archaeologists recognized that the marks were writing, and eventually translated them, because they already had some awareness of this larger context—that is, of human life and its affairs. It is our own larger context that we need to get right if we are to 'translate' our own lives correctly. Human bodies and behaviours have meanings which look nothing like human bodies—meanings to do with faith, hope, and love.

If we look around us, there is a lot to see. There is a dizzying wealth of detail, and it is hard to know what to make of it all. Gazing at the stars, or at the laws of nature, or at biological evolution, gives an impression of beauty but also the realization that this scenery is not our true context, because what we see is vast and wonderful but mindless and impersonal. Living alongside one another, we find personal interactions, which call upon all our personal resources, and we realize that our own context, and therefore our own meaning, is much more to do with this vibrant community of people than it is to do with the rocks or the stars or the evolutionary processes that shaped us. To get at our meaning we have to 'translate' with this context in mind, and it is not a question of looking *at* anything, but looking *with* or *in*—looking *with* the attitudes we learn from the wisest among us, and immersing ourselves *in* the role we are called upon to play.

10.2.1 Cogency

Now I want to return to the point I made at the outset of this section, about scientific explanations, and look at the aspect of *cogency*. Someone might accept that science has to operate on the basis of an intuition or faith that our reasoning faculties are a good guide, and they might accept that science cannot tell us either our meaning or our roots. They might find it easy to agree that atheism is not guaranteed to be correct, but they could still consider that it is not just a good bet but really the only option, because they might feel, quite strongly, that appealing to the transcendent as a locus of any sort of concern, judgement, or empathy doesn't make any sense, because simpler explanations—ones which do not unnecessarily multiply hypotheses—are always better ('Occam's razor').

This is the question with which both this and the next few chapters are concerned: the question of superfluity and lack of explanatory power. The answer involves two points. The first is that analysis does not in any case answer all types of question, and the second is the crucial fact that one simply cannot invoke or deal with God in passive language.

These two points are demonstrated by what happens when you try to investigate a person. As long as you are still analysing you can't engage with them on a personal level—you will see, hear, and touch a living body and brain, but you will not meet a person. Aesthetic experience is similar. It is not that you may not or should not analyse a painting or a piece of music or a moment of pure beauty in a football match, but in the moment when you are engaged in such analysis, then the aesthetic experience itself is not available to you. And beauty cannot be put to explanatory use—it is not a hypothesis but something that can only be seen and respected or responded to. A good argument can be beautiful, but the beauty is a property of the argument as a whole, and cannot be seen until you have first followed the argument, so it contributes at the end of the process, not the beginning, and then it goes back and lights up everything.

And furthermore, analytical methods—methods which break things down into manageable impersonal ingredients which can be scrutinized and manipulated—are not sufficient to allow us to live life. If you try to find a friend that way, you will stay lonely or immature or both. You cannot assess everything by means of a scientific treatise, because the area of *human value* is not just hard to analyse but resists analysis. There is no answer to the question 'which of your children is more valuable?' other than 'get out of here; I spit on your survey; I detest your calculator.' The very language of analysis is inappropriate, because it devalues everyone concerned.

Much of life consists in allocating our finite resources of time and energy to other people whose needs cannot all be met by us. We could try to do that by following an abstract calculation, but that would amount to volunteering to become a robot. Instead we have to find a way to do it in a more human way—playing our part in a larger life, loving other people and ourselves as best we can, relying on them to accept that we are doing our best in a situation which we do not control.

So analytical methods are not always the right tool to use, and as we approach each other, they are less and less the right tool. Therefore it should not surprise us to find that the approach to the transcendent does not fit into that sort of language. This is especially so when we are concerned with that overarching reality which addresses us as people. God is indeed not, and should never be considered to be, part of a scientific explanation. God is not *used*, but encountered and responded to.

Anyone who thinks that scientific explanations are the whole of human experience, or the only thing that matters, will, of course, end up atheist if they are consistent. However, in fact scientific 'explanations' on their own are just lines of reasoning dangling in the void; they are neither the whole of human experience, nor the only thing that matters. Without love, they are like so much noise. And it has been said, in an attempt at truth-speaking as frank and honest as any other attempt, that God is love.

When we try to say what it is we are talking about when we name God in this way, the best we can do is commit ourselves to efforts at human emancipation and, more generally, good stewardship of the whole ecosystem. This is because a phrase such as 'God is love', when spoken by us, only has the content that we manage to give it. We give it content when we do some practical politics, such as abandoning imperialism and extending voting rights. When we fail in this, both the first and the last word in the phrase are being abused. But the phrase helps nonetheless, because at least it orients us in the right direction of travel.

The sense of beauty, or cogency, that God brings, lighting up the whole argument, as it were, is found by bringing together the whole of our experience. There is no single experience that reveals God's nature, and no experience that cannot be interpreted another way, but one can begin by noticing that there is something in common between truth, beauty, and goodness—they each make demands on us, and also fulfill us, and also leave us thirsty for more. Our truth-recognizing apparatus turns out to require our beauty-recognizing apparatus, and goodness has to underpin both. No part of our truth-recognizing apparatus should be abandoned; all must come in. Philosophical, mathematical, and physical analyses make a start, but also have to rest content with their role of informing but not dictating our responses to the world and each other. We encounter each other, and we encounter ugliness and beauty. Beauty opens a window onto a mysterious and ravishing realm which eludes us in the moment we reach to hold it. Goodness in others inspires and humbles us. Cruelty and injustice is more and more objectionable. Through *all* these things we can also encounter reality itself, and reality itself turns out to be not It but Thou. Not something that I assess, but Someone who assesses me.

what we encounter in silence, and try to live up to in life, is capable of intimacy and shared endeavour. This is the sense in which God is personal.

This sense of a *personal* character to the truth which forms all things is not a piece of anthropomorphic fantasy. It is, rather, a way of putting into words two things. First, a notable aspect of human experience, and second, a sense for what most profoundly enables people to grow up. On the first: it is a widespread, and also sane, peace-loving, and natural, experience that what we encounter in silence, and try to live up to in life, is capable of intimacy and shared endeavour. This is the sense in which God is personal. On the second: connected with this is the considered view, hard to put into words, that in order to grow up we somehow need to cease regarding ourselves as the judge of all things, and yet we wish to retain a keen sense of truth and justice in the world that we are part of. To throw off the former while retaining the latter is one of the great liberations which God holds out.

Shamayim

The bedrock on which the tread of the universe gains traction;
The kiln which fired up the Dirac field;
The loom whose threads run through the wings of a fly;
The courtroom where truth prevails.

The garden in which the universe bloomed;
Mathematical needlepoint exploding into everywhere and
    when;
The fidelity of the natural world;
The fidelity to the natural world.

Helices of massive spheres in motion;
The endless seeking, seeking of the ant;
The end of pain which otherwise has no end;
The place where, if you knock, they must take you in.

A simple family meal;
Passing the butter;
Grace;
Laughter.

11
What Must Be Embraced, Not Derived

> Values and moral judgement are considered. The main business is to present a philosophical argument to show that the tools of logic—of reasoning from axioms—cannot show us what is valuable or good. No proof can show either that there are values that we should respect, nor that there are not. But to choose to deny the possibility of an objective standard is to choose isolation over exploration. And what people value highest is not well captured in terms of purely impersonal abstractions. This is a pointer towards the journey beyond atheism. The interplay of reason and faith is then discussed.

> Thinking is more interesting than knowing, but less interesting than looking.
>
> Johann Wolfgang von Goethe

As I have begun to explain in the previous chapter, the trajectory of this book is moving from a discussion of the structure of science towards a discussion of matters that lie beyond the competence of science; matters such as *values* and the choices we have to make about how and where to devote our time and energy. In short: how to be human. I expect that some readers may be broadly familiar with science, but unfamiliar with the subtlety that arises as soon as one begins to speak about moral value or duty or things that have a claim on our allegiance. Scientists usually appreciate that they have some sort of duty to honesty in their professional work, but they may never have taken time to explore the basis of this commitment. It is a basis that is not itself amenable to scientific analysis. That is what I will show in this chapter.

The present chapter presents an argument that is primarily philosophical in nature. It is an argument about moral value, and about the nature of the territory one is in when approaching anything that makes

Science and Humanity. Andrew Steane, Oxford University Press (2018).
© Andrew Steane. DOI: 10.1093/oso/9780198824589.001.0001

the sorts of claims on us that moral value makes. It can be unsettling to realize that in this territory one is not altogether in charge of what happens. The meaning of that statement will become clearer as we go along.

The discussion includes a notion that tends to raise suspicion nowadays, because it can be abused and has been abused in the past. This is the notion that values can be objective and absolute. People are suspicious of that notion for two reasons. First, the only values that humans ever get clearly in view are ones that humans *can* get clearly in view, and it follows inexorably from this that we can only ever get imperfect and incomplete knowledge of anything. Second, the notion of absolute values is abused when a person or group in a position of power claims that others ought to accept whatever that person or group asserts about absolute values. However, it does not follow from the first difficulty that humans are utterly clueless about value, and it does not follow from the second that humans are always abusive.

In this chapter I am going to present some philosophical reflection about the nature of this territory. I am not going to make a case for one value or another. I am also not going to claim to know whether some values are objective, but I will show what sort of thing follows if there are objective values.

For the sake of clarity I will first make a point about values by the following example. For one reason or another, certain moral stances are widely considered by human beings to be important and valid guides. For example, most of us consider that we should negotiate moderate differences in a peaceable manner. We do not encourage people to first arm themselves with machine guns and then turn every disagreement into a struggle to the death. In view of this, a wide and lengthy tradition of thought has been inclined to say that there is some sort of objective moral imperative that weighs in against indiscriminate slaughter and promotes considerate behaviour. What I want to do here at the outset is not adjudicate whether or not there is such an objective moral imperative, but to say what sort of thing it would be if there were. It would not be a simplistic assessment of behaviour of the kind that can be captured by a list of rules. But it could be a nuanced assessment which, though not captured by rules, can nevertheless be illustrated in broad brush or in certain respects by rules. Even someone very uneasy about the notion of absolute morality would probably not think that indiscriminate slaughter is a good thing, so that a rule of thumb that says 'do not resort to indiscriminate slaughter' should not be regarded as some sort of imposition or restriction on human liberty. Nor should the notion of indiscriminate slaughter be regarded as just another amoral and neutral option on a par with more welcome things such as efforts to promote universal literacy. I don't wish to imply that the subtleties of moral philosophy are ignored by people who think there are no absolute or objective values; my point is that those subtleties are not ignored by people who think that there *are* absolute values either.

11.1 A Philosophical Investigation

The essay of this chapter was prompted by a phrase in a sermon by Rowan Williams, in which he points out that sometimes the craving to know the answer to the question 'is there or is there not a transcendent creator?' can become a distraction, distracting us from other duties, or it can become '*an attempt to find an argument that can do work that only you can do*'. In this essay I will unpack and reflect on the truism that is contained in that pithy and eloquent phrase.

Suppose there is something that can correctly be named as 'that which deserves the deepest allegiance of all free agents'. I am using the word 'something' in a deliberately loose way here; the 'something' could be a written constitution, or a person, or a set of abstract principles about what is good, or something else. And by 'free agent' I mean some sort of responsive agent who has the capacity to be aware of the something (perhaps at first imperfectly, but with an awareness that can grow), and whose innate nature affords that agent the freedom to desire to act in accordance with the allegiance we are talking about, and who therefore so acts when circumstances allow. By 'deserving allegiance' I mean that the efforts and faculties of free agents are appropriately and well directed when they are directed to discovering the nature of and promoting the interests of that which deserves allegiance. Finally, by the words 'there is something' I mean 'there exists in reality something' and by 'exists in reality' I mean the following.

To 'exist in reality' does not necessarily involve 'being composed of the stuff of the physical world—the particles and fields that the free agents find their world to be made of', but it does involve 'influencing the stuff of the world in which the free agents live, in a coherent, recognizable way.' That influence might be quite subtle. It could include, for example, the sort of influence that music has, so music exists in reality (and I mean by this an existence that includes the subtleties of aesthetics, not merely the sound vibrations). It could also include something more subtle. It could include so arranging the processes of the physical world that they confer on the free agents the ability to recognize and respond to goodness and beauty, should the free agents so choose. It could include something more subtle still, like simply but permanently presenting the possibility of greatness, whatever greatness may be. It could include, also, a singular influence at a particular time and place, or a singular influence on one particular thread of events, which showed that the source of that influence exists in reality. The influence might take the form of a teaching and training programme, for example, carried out by hints and nudges that respect the integrity of the taught; an influence which made a new coat from an old by adding to it thread by thread. But to be said to 'exist in reality', according to the definition I wish to use, must involve the attribute of *sense-making*. It must involve that a community of free agents can come to agreement that it *makes sense* to affirm the existence in reality of whatever is said to exist. So, for example, such a community need not, and should not, affirm the existence in reality of an otherwise ordinary human who can fly, even though this is a common feature of dream sequences. However, they can, and should, affirm the existence in reality of whatever it is that makes compass needles swing due north (i.e. the magnetic field of the Earth). And, according to the definition, they can also affirm the existence in reality of other sorts of things, such as human rights. That is, the right to free speech makes sense, and to assert it as a universal right also makes sense (I mean here the right properly understood, i.e. with cautions about giving space for others to affirm their right, and safeguards around inciting unjustified violence, and that a right for one implies and is implied by obligations on others not to thwart the right, etc.). Such a right continues to be a right no matter what the external circumstances are. So the right to free speech exists even in totalitarian regimes which do not respect it and which try to suppress it—in such a regime the *legal* right does not exist (according to the definition of 'legal' adopted by the regime), but the human right does, and this fact can be correctly and honourably asserted by citizens.

The above statement about human rights is not necessary to the argument of the present essay—nothing will logically turn on it. For present purposes it serves merely to illustrate the breadth of what I am referring to when I use the concept 'to exist'. It would be possible to deny that human rights meet the criteria that I am employing for 'exist' (namely, to make sense and to exert an influence), and still accept the main argument of the essay.

The phrase that I have put forward above ('that which deserves the deepest allegiance ...') is, admittedly, too brief to deal adequately with the subtleties of what allegiance can involve. One may have several values or principles one wishes to support, and there are often circumstances in which these principles are in tension with one another, and it is not clear what form allegiance to all of them should take. This may make one feel that there is no such thing as a deepest allegiance. But in practice what we do in situations like this is to do our best to make a fair response to the tension, admitting the impossibility of getting all desirable outcomes. So the thing that recommends this kind of reaction has our deepest allegiance.

The remarks in the previous paragraph should also resolve another issue. One might want to say that 'that which deserves the deepest allegiance' is a plurality, not a singular. I have some sympathy with that reservation and do not want it to hinder the argument, because I think nothing essential turns on it. So allow me to convert this plurality into a singular, by collecting it together. The sorts of things this plurality might have in it include reason and creative imagination, but the point is that it does not include everything; it does not include unreason and refusal to explore, for example. Or so I assert. Therefore it can be collected together and we have a collection that does not include everything. Such a collection of valuable items, including the guidelines or generous instincts that teach us how to handle tensions among them, can be considered to be that which deserves our allegiance.

So much for the definition of terms. Now it will help us, for the purposes of the present argument, if we have a shorter name that signifies 'that which deserves the deepest allegiance of all free agents'. I will use the name *Standard*. This name has been chosen to suggest the definition, and to leave open what else may be found to be true of the Standard.

First I would like to comment on what we are considering. On supposing that there is a Standard, we are making a supposition about the universal brotherhood and sisterhood of all free agents. That is, we are saying that free agents are called upon to recognize some, at least, of what we may call *shared values*. The value of teaching over coercion, for example, and the value of justice and mercy over revenge, and the value of efforts to provide conditions that allow another to prosper, over efforts to suppress or kill another.

Now the question arises, do we have to merely hazard a guess that there is this Standard, or can the existence of the Standard be derived by a process of reasoned argument?

Let us call this question Q. I wish to show that the question Q is somewhat confused, and it does not offer all the options.

First, Q does not offer all the options because, in the absence of a derivation by reasoned argument, which I will unfold below, there are more options available to us than a merely haphazard guess. Rather, we can observe what is going on, listen to people whose actions command respect, and we can form an opinion which is reasonable though not formally derived, and which is recommended by various considerations. The considerations will include how it makes people behave, and how it influences us when we try to live by it. One can bring in many further reasons to accept that the Standard is real, and one can show that such an acceptance is not irrational, but one cannot place oneself or anyone else under an obligation to accept it as the outcome of a logical proof. This is the main point of this essay, to be presented in what follows. After all the weighing of considerations, the only action available is to bet your life, because that is all you can do with deepest allegiances.

Next we address the confused nature of the question Q. We will demonstrate the self-contradictory nature of what it offers as a second option (the derivation by a process of reasoned argument).

Contradictions in reasoning processes can happen in several ways. For example, suppose one set out to prove that there exists a pair of integers p and q, both perfect squares, whose ratio is equal to 2. The proof will fail because in fact there does not exist such a pair of integers. In this case the existence cannot be proved because the existence does not hold. However, proofs can fail in more subtle ways. Gödel famously showed that systems of logic and of mathematics always exhibit statements or theorems whose truth value cannot be ascertained within the system. This does not in the least imply that such statements are in fact false. They may equally be true. But their truth or otherwise is not demonstrable on the basis of the axioms and methods of the logical structure in question. A similar type of undecidability can occur in the realm of existence claims. It is not hard to construct examples. Consider an item X which has the property, 'X is such that the existence or non-existence of X cannot be proved by a process of reasoning from axioms'.

What Must Be Embraced, Not Derived

In this assertion it is taken for granted that the axioms do not include a direct statement about the existence or non-existence of X. Suppose that one wishes to discover whether or not there exists in reality such an X. Maybe there is such an X, maybe there is not. If there is, then that existence will not be provable, because if it were, then we would have a contradiction. If there is not, then that non-existence would also not be provable, because if it were, then the thing whose non-existence had been proved would not be X.

We will now show that the Standard has the property we just considered. We will show that the Standard is such that the existence or non-existence of the Standard cannot be proved by a process of reasoning from axioms.

Any attempt to derive the existence of the Standard by a process of reasoned argument must first assert two things. First, the axioms adopted for the purposes of the argument are asserted. Second, it is also affirmed that *reasoned argument* is itself valid. In making the argument, or in inviting others to accept the argument, we must first assert that reasoned argument is a good and trustworthy activity, and invite others to agree with that assertion. Now please make no mistake, I fully agree with that assertion, which you can see because I am engaged in reasoned argument right now. But I recognize it as a truism whose truth cannot be derived by reasoned argument.* Rather, it is lived by and found to be a sound guide in that way. A community of free agents that accepts it (i.e. the value of reason) finds in practice that things go better than if they do not accept it. Somehow or other, that is the way the world has been configured.

The same impossibility of derivation applies to the Standard. For we described the Standard as that which deserves our *deepest* allegiance. But any attempt to derive the existence of this Standard, as if we were unsure and needed the derivation in order to place our allegiance on a firmer footing, is thereby asserting that something other than the Standard has our deepest allegiance. We would be saying 'that which I am first and foremost willing to trust is not the Standard, but the axioms and method of such-and-such a reasoned argument. I trust those axioms and that method first, and use them to provide me with a reliable guide—a guide to the existence of the Standard, for example.' Any such

---

* The difficulty of deriving by logic the rules of logic was memorably discussed in a famous article by Lewis Carroll, called *What the Tortoise said to Achilles*.[1]

statement is self-contradictory. I think the above already strongly hints at this self-contradiction, but further argument is required to place this assessment on a firmer footing.

I propose to show that it is self-contradictory to start out from something other than the Standard as foundation of any argument for the existence of the Standard. One can sense that there is a certain tension in this situation, but it is not a trivial problem to show precisely where the tension lies and that we have a genuine contradiction.

Let A be the supposed existence argument. We can prove that A must be invalid by showing that a self-contradictory situation would follow if it were valid (*reductio ad absurdum*). The self-contradictory situation is that of a rational free agent who is supposed to be in possession of A. Such an agent can reason as follows.

1. I do not in the first instance consider it my rational duty to either affirm or deny the existence of anything or anyone in the role of the Standard.
2. However, I have argument A, and on the basis of A, I arrive at affirming the existence of X, where this X is none other than that which deserves the deepest allegiance of all free agents.

Now consider the following cases, which exhaust all possibilities: either $A = X$ or $A \neq X$. In the case where $A = X$, the existence of X has not been derived from other considerations, but asserted at the outset. In the case where $A \neq X$, anyone who claims that A is in fact the Standard will finish in a contradiction. Hence A is not the Standard. But, according to A, X is the Standard and exists in reality. Therefore allegiance is correctly given to X before one considers whether or not A may be sound, since this is the response which, by definition, is correctly and appropriately given to that which deserves the deepest allegiance of all free agents. It follows that the relationship between X and A is not that the existence of X is derived from A, but rather, the validity or otherwise of A is assessed by enquiring whether or not A is consistent with loyalty to X. In other words, the statement 'A is valid' can only be true if the character of the Standard is consistent with the validity of A. However, the statement 'The Standard affirms the validity of A' implies the truth of the statement 'recognition of, and hence loyalty to, the Standard does indeed follow from A', which in turn implies the truth of the statement 'Argument A can itself serve as the starting point from which all other loyalties appropriately follow', which is equivalent to the assertion 'A = Standard' and therefore A = X, and we have a contradiction.

In summary, anything whose existence can be validly derived from other considerations is, by the very fact of that derivability, not something which deserves one's allegiance before all else.

Let me deal with a possible objection at this point.*

Objection: This argument is unsatisfactory. A Standard that demanded ultimate allegiance would not preclude there being a proof of the Standard's existence. The axioms and reasoning could be trusted to show the existence of that Standard. Whether the person providing the proof then followed the demands of the Standard is another matter. It that person did, then they would have to abandon the proof. But the proof would still exist.

Reply: I have some sympathy with this objection, because it does capture what often happens in human life. A person may arrive at a destination having followed a certain line of argument, and then, having arrived, decide that the argument was not adequate to support the commitment they now propose to give to the destination. This happens in science, for example, when one arrives at a new way of seeing things and it is the new way which itself commends itself, even though the existing experimental support may be limited. Think of the discovery of general relativity, for example, and the ambiguity in the fossil evidence at the time when Darwin published his *Origin of Species*. However, as a matter of logic, I stand by my argument. The point is that no supposed existence proof for the Standard can hold up under scrutiny, if the proof is based purely on the type of tools that logic can provide, because of the contradiction I have showed. The situation is very similar to the impossibility of proving the consistency of the rules of arithmetic by using the tools of mathematical logic (Gödel's work). It does not follow that one should distrust arithmetic. It also does not follow that one should be leery of acknowledging and indeed celebrating a Standard.

We have obtained a contradiction both in the case of any argument supposed to prove the existence of an X deserving allegiance *greater than* or *prior to* that appropriately given to the axioms and to reasoning itself, and also in the case of an X supposed to be *equally* reliable or trustworthy as the axioms and reasoning. For if the argument held good then in

---

* Thanks to Dr P. Marriott for raising it.

either case the X would be derivable and therefore could reasonably be placed at a secondary or subordinate position, as acquiring the proper allegiance of free agents thanks to their prior allegiance to the axioms and reasoning. So once again, as soon as such an X is derived it is not X, and we have a contradiction.

So far we showed that the existence of the Standard is not provable (because not the kind of thing that can be proved). Next we show that the non-existence is also not provable.

Suppose one furnished an argument which purports to prove that there does not exist that which deserves the deepest allegiance of all free agents. When one free agent shows such a 'proof' to another, the second always has the option of rejecting the axioms or process of deduction, or both, on the grounds of inconsistency with a prior commitment to the Standard. We have here an unusual sort of situation in logic. The first free agent may feel convinced of the correctness of her axioms and reasoning, yet, logically, she must accept that the second agent is being equally consistent and rational when he rejects the argument on the basis of his prior commitment to the Standard. So then the first agent has furnished a sequence of statements which may be said to amount to a 'proof', yet it is a 'proof' which every agent can rationally reject if he/she so chooses. Therefore it does not amount to an example of what the word 'proof' signifies. The conclusion is that the non-existence of the Standard is not provable.

To summarize, it is not possible to prove (in the sense of deduce by logic from axioms) either the existence or the non-existence of that which deserves the deepest allegiance of all free agents. This is not owing to some sort of decision to be elusive, nor to a failure on anyone's part, but is simply and directly owing to the innate and logically unavoidable character of the situation in question. The Standard is not making an unreasonable demand—a demand for completely unmotivated allegiance, for example—but the Standard is what the Standard cannot avoid being. I already explained that this does not mean one is left with no access to reasonable means to learn of the existence and nature of such a Standard.

Now we discuss examples and implications. One clear implication is that it is not possible for any free agent to insist that another must accept the first's opinion about the nature and existence or otherwise of the Standard, by appeal to a proof in the form of an irrefutable rational argument. On this sort of question, rational agents can only recommend their opinions to one another in other ways.

Another implication is that it is self-contradictory to assert the value of *reason itself* over and above the value of the Standard. One trying to do this is really just placing 'reason' (suitably defined) in the role of Standard. But an alternative (and in my opinion wiser and altogether more satisfactory) position is to assert the value of reason by saying that the Standard is such that loyalty to the Standard includes affirming the value of reason.

Of course one might instead assert the non-existence of anything matching the description I gave for the Standard: 'that which deserves the deepest allegiance of all free agents'. One might assert that there does not exist such a Standard. Such a non-existence assertion amounts to the announcement of the non-brotherhood and non-sisterhood of all free agents. It says there is no basic set of values that they should have in common, and nowhere outside themselves to seek a better assessment of how they are getting on. Some can affirm reason, empathy, and creative imagination working together, others can affirm unreason and repetitive mantras; some can affirm nurture, some can affirm torture, and that is all that one can say, because there is no Standard, nothing that deserves the deepest allegiance of all free agents.

I do not recommend this fragmentary, inward-looking, and lonely possibility. But perhaps I have overstated this conclusion. Perhaps this is not the conclusion that would follow. Because if there is no Standard that deserves the allegiance of *all* free agents, there might be one that deserves the allegiance of all human beings, at least, or of some other group. The trouble with this is that as soon as one begins to divide up the free agents in this way—some who should respect the Standard, and some who have no such obligation—then one opens the way to some laying claim to the right to lord it over others. If the Standard affirms that people should not lord it over each other (as I believe) then the people respecting this obligation will refrain from that behaviour, but the people who genuinely have no such obligation genuinely can do as seems good to them, and there is no sense in which their choices will be either right or wrong. Their freedom is illusory: it is the 'freedom' of meaninglessness, which is no freedom at all. And meanwhile their victims suffer for it.

So then let's try the next possibility, and suppose there is nothing that can properly command the deepest allegiance of anyone, or of any pair. I can't see how this can make the situation anything other than worse. Logically, this may be the situation we are in; I am not claiming to know. I am expressing the feeling that I hope it is not like that, and the willingness to act as if it is not.

Someone may object that we don't need a Standard in order to come to have shared values; we just come to one another and find by negotiation and so on that we arrive at plenty of shared values. The problem is that what is being agreed upon in such meetings does not merit the name 'value'; it is merely 'opinion' or 'what we like', rather than some notion of discovery or having something to live up to. It leads to a bleak picture of human identity: lots of individuals coming to opinions and agreeing or not as the case may be, with no prospect of genuine learning available, because there is nothing there to learn.

To illustrate this, consider what happens when a friend or a family member gets caught up in an extreme religious sect or in a terrorist movement. The friend honestly considers that the best thing he can aspire to is to deliver up his money and sense of self-worth towards some guru figure; or the best thing he can do is try to spread terror among third parties who bear him no ill will. Those are his 'values'; that is what he thinks he owes allegiance to. If there is in fact no way we can appeal to a value which he and we ought to hold in common, if our values are just our opinions, then if he will not be persuaded to change his mind, we sense that he is drifting off, and he becomes severed from our sense of mutual belonging. There is nothing for him to learn other than that our opinions differ from his. If, on the other hand, we can say that, no matter what we think or our friend thinks, we are all called upon by the same Standard, then we can never give up trying to come to a shared vision. Note, it is not a case of claiming to own the Standard nor claiming unique authority to discern that Standard. It is certainly not about making oneself the judge over another. It is simply a case of refusing to completely abandon one another to whatever our delusions may be.

This is an extreme case, but the sense of drifting apart begins, I contend, more gradually, as soon as we choose to consider that values are purely subjective. It is better—more hopeful, more tending to learning and the increase of fairness and joy—if we consider that value is there before and beyond us, to be learned, appreciated, and grappled with more and more fully.

If we explore this, then the next possibility is to recognize the Standard and affirm that it is impersonal; it is something like 'natural justice' or a collection of principles such as 'liberty, equality, fraternity, reason, goodness, beauty, imagination, truth'. This is one possible version of the position called 'atheism'. An important movement within atheism takes place when we see these lovely qualities embodied and lived out, in a human being and a human life, for example, and then many of us feel that something even more remarkable than the qualities themselves (as abstractions) is on show. So we are inclined to begin thinking that the Standard is not the impersonal qualities alone, but those qualities embodied and lived by. Now the Standard is getting a half-way-to-personhood sort of attribute. That which deserves our deepest allegiance becomes something along the lines of *participation in a community of free agents living in love and truth*. Also, we begin to see those free agents as themselves reflections or embodiments of aspects of the Standard.

Finally, one may ask: is, then, the Standard perhaps something not quite so impersonal as we had guessed? Might not *that which deserves our deepest allegiance* have something of the quality of a community of free agents living in love and truth? I mean, here, not necessarily something already expressed in physical entities, but that which those physical entities are trying to reflect? Also, one may ask, what does promote that very attempt? How does a community of free agents living in love and truth come to be realized? To a surprisingly large degree, it comes about by a combination of two insights on the part of the free agents: first, that they are valuable and lovable, and second that they have much to learn and must endeavour to do better. So that which deserves our deepest allegiance includes these insights, and it includes that which makes these insights really come home and take root in the hearts and minds of the free agents. And we see, imperfectly and tentatively, that sense can be made of all this by allowing the admissibility of a step beyond atheism. A humble and courageous step.

11.2 Reason and Faith

The previous section is self-contained. What I called a philosophical investigation or essay finished there. The present section will present a related issue, which is the age-old tussle that is often charaterized as one between reason and faith. It is my aim here not to show which side 'wins', but to show what the nature of this tussle is, and therefore why it cannot be 'won'. This is because it is closely related to the issue discussed in the previous section.

The word 'faith' has for a long time had connotations of trust and loyalty. For a long time it referred to that aspect of human behaviour which involves willingness to go in a certain direction, and to trust, motivated by a sense of value combined with a reasonable assessment of suggestive evidence, but not coerced or unavoidable. However, more recently that same word 'faith' has been widely used to mean 'beliefs arrived at without any basis in evidence'. It is only the former meaning that interests me here. (The second and more recent supposed 'meaning' seems to me to be a more-or-less political attempt to change the meaning of a word; it is a nasty and narrow-minded attempt to redefine the language of your intellectual opponents.)

In the following I will consider faith and reason together, with a view to showing how they work off one another. As I have said, faith is essentially a kind of willingness combined with a sense of value. Reason is about being receptive to persuasion, and honest enough to follow a sequence of steps where the connections can be shown and seen.

Faith involves trust. So why don't we just say 'trust'? Why talk about 'faith'? It is because of a subtlety that arises whenever anyone tries to interpret what is going on around them. What is going on includes brute facts about the displacement of people's physical bodies, and the connections made between neurons in their brains, but those physical facts do not show us the meaning; we have to interpret. Is the meaning simply the proliferation of assertiveness? The strong/quick/confident 'win'? But what is it to 'win'? Is it to have more money, or to have more courage? 'Life is short. Have an affair', says the Ashley Madison website. 'Life is short. Keep your integrity', we might reply. This is where faith comes in. This doesn't necessarily mean faith in a trustworthy absolute reality which relates directly to persons. It could be faith in atheist philosophy as a worthy way of 'reading' the world and consequently living in it. It is faith because one is trusting some set of notions to be worthy, to be worth something, valuable. One feels it is so, and one's reason concurs.

One of the ways in which this is subtle is that, in the end, you cannot sit in judgement on your own ultimate values. You can only throw yourself into them. This is the upshot of the philosophical argument I set out in the previous section.

What Must Be Embraced, Not Derived

Consider for example something that most of us value, such as the principle of equal opportunity. Do I affirm the goodness of this principle simply because it is, in itself, something that ought to be supported, which I can recognize, or do I affirm it because I myself derived, by a process of reasoning, that it is good?

Let's consider the second possibility: that one could somehow deduce the goodness of equal opportunity, by some sort of reasoned argument. How could anyone deduce that? Only by connecting it to other things whose goodness they affirmed, such as the promotion of human happiness. So ultimately this process has to finish in something that we do not stand in judgement over, but which we simply recognize as worthy of our commitment. It follows that one *cannot* be the type of human being promoted in some rationalist circles, one who stands like a little god at the centre of his/her world, assessing everything and deciding what is good or bad. We can only commit ourselves to values and duties that we did not announce, but which announced themselves to us, and called upon our allegiance. Human life consists largely in submitting ourselves to a welcome duty that we feel ourselves to be under.

What Jewish and Christian* theism is all about, ultimately, is that in such a commitment we are committing ourselves not to an impersonal abstraction but to that which calls upon us in love and compassion and solidarity. We have a lot of reasons to think that, but I am here expounding why it is that faith is nevertheless a central part of the package.

There have been long theological debates over the roles of reason and faith, especially in the medieval period in Europe. These were about whether faith or reason ultimately carries the weight, or does the primary work of what opens us up to acknowledging Godness, not just obeying goodness. It is the echo of those debates which modern-day atheists sometimes detect. You can find quotations from people such as Martin Luther which, taken out of the context of the issue they were addressed to, appear to disparage reason in general. So now people read the history as if religious people undervalue reason and assert faith as an alternative. But in fact those debates were more subtle than this.

What the essay of the previous section shows, and what philosophers and theologians have realized, is that the area of ultimate values is one

---

* I mention these two because they are closely related and with them I know sufficiently what I am talking about.

which simply cannot be navigated by reason alone. It is just not that kind of area. The process of reasoned argument cannot get anyone all the way to a statement of what is real and good, because that is the very nature of the case. A rational demonstration of what is good, or of the nature of goodness, cannot be based on anything other than some sort of standard of good as a starting point. It follows that if there is an objective or absolute good, which I shall here label Y, then in the end, what opens us to Y, or brings us reliably to Y, can only be Y Yself, not our own efforts at deduction (nor anything good we may do, for that matter). We can only trust ourselves to the process. *What reason shows us is that this is the nature of the case.*

The reasoning I just described does not show us that this absolute goodness will be capable of loving purpose, of being One Who has opinions, rather than an impersonal 'It'. But what reason can and does do for us is the following. By a process of reasoning we deduce that we can only discover the nature of true goodness by allowing such goodness to declare that nature to us. All we can bring is the willingness to somehow position ourselves where we can see or be shown the nature of good.

It is not that there is no evidence to consider. We are not expected just to launch ourselves off into we know not what (although that can be a rather humane, truth-seeking, and open-hearted thing to do, as long as the 'we know not what' is signposted by the very highest virtues one can think of). Rather, we can listen to each other, and to the rest of the natural world, and to the wonders of science, and to the exquisite yearning that beauty creates, and we can assert the horrible objectionableness of meaningless suffering, and we can see what the most impressive lives have to show us. One of the turning points in human history occurred when one of the most 'impressive' people identified so strongly with the so-called nobodies and the unjustly accused that he managed, for a while, to overturn the pernicious idea of a heirarchy of human worth. Suddenly people had their eyes opened to each other. They began to see everyone as equally valuable, and thus was born the view of the individual person as valuable in and of themselves, rather than as an instrument of family or society. This is the conception that human emancipation has built upon ever since, and which we now often take for granted. Of course the idea of a stratified society reasserted itself, but less strongly, and now the possibility that it need not be so was out in the open.

Here is Not

'Show me.'

Here is not that which you see
but that by which you see.

'Let me have.'

Here is not that which you hold
but that which holds you.

'Let me consider.'

Here is not that which you assess
but that which assesses you.

'Why?'

Here is not that which explains
but that which bears.

'I'

Here is not that which you give
but that which gives you.

'___'

Here.

12

Religious Language

> Religion is considered as a social phenomenon, having both good forms and bad forms. The widespread modern nervousness around religion is recognized, but that nervousness itself often misconstrues religion. Religious violence is briefly analysed, and compared with other forms of violence. The variety of ideas that people mean by the word 'God' is sketched. The main aim is to point out that there is this variety, and to say that religious language has to be indirect.

In the previous chapter I was careful not to use the word 'God', but towards the end I was obviously exploring territory close to what good religion is largely about. In subsequent chapters I will be addressing further issues that are in this territory, but I am conscious of writing for people who may be uneasy about religion in general, having serious reservations about the intellectual and moral credentials of things that religion often seems to take for granted. Religion often seems to be a strange mix of half-thought-through hand-me-down assertions about untestable propositions; odd ways of speaking and behaving; the bizarre suggestion that certain plainly imperfect books are in fact perfect; the framing of morality in immature ways such as rules, rewards, and threats; uncritical thinking; and refusal to face evidence. I say 'seems to be' rather than 'is' here, because although all those characteristics are to be found in religion (because religion is difficult and is practised by faulty humans) it is debatable whether they are its defining characteristics. Religion is largely about the peaceful aspirations of ordinary communities to celebrate life in the round, with an emphasis on gratitude, mercy, giving, scholarship, courage, honesty, hope, and wisdom. Really. That is what is mostly going on in all those communities that form themselves around churches, temples, synagogues, mosques, and the like. However, religion is also dangerous, as fire and sex and tectonic plates and many other good things are dangerous. And, obviously, it is open to abuse.

Science and Humanity. Andrew Steane, Oxford University Press (2018).
© Andrew Steane. DOI: 10.1093/oso/9780198824589.001.0001

At the time of writing, militant groups in several parts of the world, including Kenya, Nigeria, and the Middle East, and lone gunmen elsewhere, are using Islam as a way of framing their murderous desires and policies. The resulting misery and ignorance and waste is largely the responsibility of the people directly producing it: the adults who formulated the thought, 'a chance to enact murderous violence without moral qualms? Yes please, sign me up'. It is also, in part, the responsibility of mullahs who told young men to arm themselves with Kalashnikovs rather than negotiate with each other; and it is in part the responsibility of manipulative leaders who lord it over their countrymen and promote narrow-minded education in schools; and it is in part the result of international politics and the emphasis it gives to protecting the position of the powerful and the comfortable of all nations.

This example of inhumanity certainly has religious entanglements—or entanglements with certain ways of being religious—but one cannot deduce from that any simple formula for the origins of violence in general in the modern world. If one made a tally of avoidable misery associated with one ideology or another over the past sixty years, then none emerge with a proud record. The dividing line between peaceful intelligence and brutal folly certainly does not fall neatly between religion and irreligion (think of the Chinese famine of 1958–62, for example, and the Khmer Rouge, and present-day global trading practices, the marketing of addiction, lead poisoning, kleptocracy, and the impact of global warming in the coming decades).

Since this book is an attempt to contribute constructively to the understanding of religion, I had to think about whether to take the option of defining myself into a neat spot where I could look down on religion as an anthropological phenomenon and declare myself free of its entrapments. I have not done that because I find that such attempts lack self-awareness, clarity of thought, and a sense of fairness. I think it better and more honest to admit that what is called 'religious' includes a variety of ways of thinking and living, some of them good, some of them bad. Equally, the club of the 'free-thinking' is just as capable of delusion and projecting a controlling or coercive atmosphere as any other group. In short, no one owns the card that gives them entitlement to the claim that they and their friends represent the true voice of reason, or of wisdom, or of love. We can only claim that we aspire to such things.

Monotheism can itself be understood in more than one way. For some it is the idea that people should see themselves as servants of an invisible unchallengeable Ruler, and get their direction from isolated texts drawn from a holy book. This kind of thinking is acutely dangerous. It gives to people a way of justifying (to themselves) their own brutal instincts, and thus opens the door to indiscriminate violence. So the job of organized religion is not to promote this dangerous attitude, but to question it. Good religion avoids hatred of the 'other' by insisting that to serve God is, *by definition*, to serve one another, in ordinary, everyday, practical ways. There is no inscrutable unchallengeable invisible entity that we have to bet on or try to please. That is not what well-established theistic language is about. Rather, we seek to become part of the expression of generous, self-forgetful goodness that the universe falteringly expresses. To join in with this is where our meaning lies; it is who we are; it is what the phrase 'child of God' refers to. The word 'God' refers to that foundational reality whose nature is continually being more fully expressed as the universe develops. The natural world is morally ambiguous precisely because it is *not-God* and is (partially) free. What we claim by faith is that in any situation it is those parts that present goodness or beauty that reflect the deeper truth and summon our cooperation, not the parts that present pointless pain or ugly self-centredness. What we are also claiming by faith, and championing by action, is that the arc of the moral universe bends towards justice, as Martin Luther King famously put it. Religious violence is connected to a completely different agenda, one which abuses and insults God in the most extreme way possible, by declaring God to be the champion of murder, torture, injustice, and the supremacy of one group of people over others.

> *The word 'God' refers to that foundational reality whose nature is continually being more fully expressed as the universe develops.*

Modern-day atheism generally takes the view that the way forward is to regard religious language as uniformly suspect and problematic, even in its most creative manifestations. The trouble with this is that it amounts to another form of totalitarianism, because it suppresses a basic human right, which is the right to acknowledge that our existence is not fully comprehended by us, but reaches towards hope and the possibility of a great goodness. It is a profound freedom that we may say this and celebrate it. We are allowed to assert that the truth of the physical universe is not contained purely in its impersonal elements. Having checked its moral and intellectual credentials, we can enjoy this freedom to the full. So religious violence has to be combated another way. This involves, among other things, getting a more accurate and less simplistic grasp of what drives it.

When people are disaffected and unable to see a better future for themselves, they are drawn to violence, and eager to find any way they can to justify this violence to themselves. If it wasn't religion then it would be something else such as nationalism or racial pride, and in fact the latter two, or things like them, are what conflict is usually about.[10] The generic phenomenon here is not religion but tribalism, and it can take many forms, including class struggle; workers' solidarity; violent political power under the emblem of liberty–equality–fraternity; dynastic political power devoted to its own security; cartels; warlords; the man-made famines of the twentieth century which appeared under the name of 'progress'; self-proclaimed rationalism and the arrival of the 'brights'; and the ascent of the 'intelligentsia' with their natural right to rule. The latter will not advocate gross violence, but clinical methods such as de-personalized working practices, questioning the mental health of those they disagree with, taking children into state care, and 'correction' institutions.

I am not going to assess the history of violence here, nor present an assessment of the role of religion in violence. I remark merely that there is a great deal of evidence on *both* sides of the account—religion has been a powerful promoter of both good and ill in human history. And, for the sake of clarity, one should add that the former part of it is not just about charitable acts by everyday people; it is also about powerful advocacy for human rights, the vision for and bringing about of such things as universal education and welfare, and also the promotion of science and scholarship in general, as well as the arts. In the history of social reforms, it is no use pointing merely to religious efforts on the wrong side of the argument, because one almost always finds religious motivations on *both* sides.[73] The issue is the effect of such motivations on the material (the human attitudes at a given time and place) on which they act. It is, in all honesty, not that hard to tell apart the generous instincts that characterize good religion from the fearful ones that characterize bad religion. In any case the effort to tell them apart should be made.

What I do want to do in this chapter is to express how much care is needed when talking about God, and to make an appeal for imaginative, positive, creative, thoughtful such talk.

In modern-day academic circles the word 'God' is mostly used in a way that dismisses from the outset any sense that the word could refer to something to which the speaker owes anything. Academic types mostly talk about 'God' as they talk about about literature or political theory—it is taken for granted that one is speaking about human efforts to marshal people and ideas. The speaker says 'God' but they do not have in mind a beautiful reality which they are called upon to live up to; they have in mind a concept (often rather primitive) that they can look down upon and steer this way and that. This way of speaking about God is considered by its practitioners to be objective and well-informed, but on reflection one must admit that it utterly fails to even begin to grapple with what the word might mean. What the word traditionally refers to is a combination of what is found and yet sought: the word refers at once to the glorious mystery which certainly surrounds us, and also to what we sense to be a transcendent good that quite properly calls upon us. But by denying such possibilities from the outset, modern academic discourse has very often so completely redefined the word 'God' as to make God disappear from speech.

In modern religious circles, people tend to the opposite extreme: they mean well, but are too quick to speak of God. They do intend to speak about the beautiful reality that is the support of physical things and the promoter of what good the world embodies, but they often forget the degree to which we are all constrained by human ignorance. Organized religion is needed not as an end in itself, but because it reminds us of our ignorance, while serving as a vehicle to preserve the best of what the community has discovered in the past.

Another important lesson that I think we should insist upon is that unquestioning veneration of a written text—any text—is dangerous and wrong. Human efforts to write about God are properly respected when their limitations are not denied or ignored but admitted and argued over.

A third basic lesson is that whenever any principle or shared thought results in people feeling a greater connection to one another, there is immediately the risk that they will begin to hold themselves apart and resist recognizing that they also have a connection to, and shared future with, others outside the group. This is part of what human relationships are, and it cannot be completely avoided. When two people marry, they cannot avoid the fact that their relationship is more intimate than the one they continue to have with their other friends. When a family adopts a child, they are bound to be more fully committed to the welfare of that child than others not in their family. The solution to these sorts of issues is, one assumes, to try to make the increase of one link not cause the reduction in another. Married people must try to make their marriage one which serves the wider community; adopting parents try not to reduce their care for the many, but simply increase their care for the one. This is difficult because we struggle to enlarge our hearts, but one can at least get the goal in focus. Religious groups should equally see their role in terms of offering a better basis for service of others, not a basis for self-satisfaction or for triumphing over others. The same can be said for groups which form around non-religious shared values.

There are any number of books in the bookshops which will tell you 'God wants this' and 'God wants that'. Sometimes it is nasty things such as 'God will send the unbeliever to hell', or, worse, 'God wants you to punish the infidel'; sometimes it is kinder things such as 'God wants you to know that you are loved'. But, whatever the message, such books leave many of us a bit queasy. I always want to say, first, 'how do you know?' and then 'well OK, but what makes you God's spokesman? Why can't God speak for Godself?'. Also, I want to say, 'why bring God into it? Why not just say "I want you to punish the infidel" or "I want you to know that you are loved"? Wouldn't that be more honest?'.

Once you begin to notice how much human assertiveness, confusion, imposition, and well-meaning invention is tied up with speech about God, you begin to realize that the only really trustworthy sources must be the less strident ones. It is the sources which are distinctly quieter, and which spend large amounts of time in doubt, or in living out their contribution in quiet acts of service, that should command attention. But this goes against large parts of what modern political practice, academic assessment exercises, and popular culture champion. It means that, unless you are very careful, the television and the Internet will mostly misinform you about this area, because it is an area that those media are not well placed to handle. There is no camera pointing at the quiet acts of service that go on, because if there were then they would not be quiet.

When some people talk about God, they have in mind a supernatural being, a powerful such being who inhabits another realm or dimension, beyond or 'outside' the physical universe. That is what some proportion of theists, and a large proportion of atheists, mean by 'God'. The first group means this and thinks this supernatural entity exists in reality; the second group means this and thinks the entity does not exist in reality but is an invention of human imagination. Both agree that you cannot touch, see, feel, hear, or otherwise directly measure this entity. Members of the first group sometimes speak as if they can hear God, but what they mean turns out to be that they interpret some of their thoughts or inclinations as from God.

When other people talk about God, they have in mind a reality not quite so easily expressed.[65, 62] They are trying to find words to speak about the human experience of *venturing out* and *being met with*; an experience in which one finds the ordinary relations of human life in the material, physical world to be not only like the walls of a house where we live, but also like a window or an eloquent gesture towards something valuable—a depth and fullness of life that the world has never yet fully expressed. Never fully expressed, and yet expressed all the time in small ways, especially in everyday acts of courage and hope in painful situations. And there is also a lovely playfulness at the heart of it; a delicious pricking of our bubbles of self-importance. Such experiences invoke in us lengthy thought, which proves to be puzzling and difficult, but the conclusion presses itself upon us: we end up declaring that, despite the mixed picture that the world presents, it is love, not its opposite, and not mere machine-like churning either, that is the be-all and end-all. The notion of 'a supernatural being' doesn't quite hack it here, not because we are embarrassed by the word 'supernatural', but because this phrase is too much about power, and too concrete. It seems to make God out to be like other things, only more powerful, and the trouble is that then 'God' ends up being just another part of the world, but with the unfortunate property of not existing; not actually being there at all. So that type of approach is atheism. But the word 'God' gets its meaning back when it refers to that which stands in relation to the physical universe differently: not an extension or an item, but the convener of meetings, and as much about hunger as about satisfaction. The rationale for the justice we are hungry for; the unquenchable love we wished we expressed. Hence, for many people, the best single-word noun or description for this not-so-easily-pinned-down reality is 'Love'; but they would want to add that they mean by this something that includes a kind of immediacy, a simple *dasein*, a form of love which is to do with existence or sheer *being* itself.

In the end this is expressed not by blather about love in tracts and lyrics (or academic books), but by patient, difficult, hard work on behalf of others with reduced opportunities. It is expressed by being it, by embodying it in honest perseverance, good-natured fun, square dealing, prison reform, debt relief, lasting homes, top-quality scientific research, artistic expression, and the like. What we need, and receive if we learn how to receive, is the capacity to live that way.

A lesson that emerged both in the Far East and in the Hebrew tradition is a profound awareness of the limitations of language and labels when it comes to speaking of what calls upon our commitment. In Eastern traditions this is associated with deliberately indirect or paradoxical language and the use of creative tension. The ancient Israelites, meanwhile, moved on from tribal religion remarkably early, and developed the healthy practice of not having any short label that you can slap on to some super-being and imagine that you have named God. The Hebrew *Tetragrammaton*, for example, which is often said to be a 'name' for God, did not function as a label (not least because they refused to say or even write it down very much). Rather, it is an open-ended pointer whose meaning can be mulled on; its meaning is something to do with 'that which is', or 'one whose nature may be discovered'. Modern-day anti-religious writing is largely ignorant of subtleties like this. When one reads a modern polemic about a 'god' called 'Yahweh' it is usually a display of marked ignorance about the learning process that good religion is largely about. If people, whether in the Bronze Age or in the modern age, imagine God as like a god, a sort of power-being, a kindly king sitting on a throne (modern), or a warrior riding a golden calf (ancient), then they are in both cases ignorant of the journey that is described in the Bible. What we have in the Bible is a record of the fact that the Hebrew conscience rejected that whole way of thinking again and again, until it became second nature to reject it. The Jewish practice of being reluctant to write a name for God, and the refusal to create images or statues, is a centrally important part of the record, because it is an effective way of insisting that God simply cannot be captured by short statements or readily defined images, nor by anything which falls short of a standard of justice, insight, wisdom, mercy, and fun which has never yet been realized in human affairs.*

---

* The *Tanakh* or Hebrew Bible is the record of a lengthy struggle to understand the notion of justice more fully, and to extend it. It is a record of widening and

Another name could be, simply, 'Real,' or 'that which is, without being derived from anything else'. It is not a guess about what caused there to be a universe, and it is not a question of whether or not some entity called 'God' exists; rather, the word refers to that which is real but not altogether known. Our role is not to derive the existence of reality but to discover and respond appropriately to the nature of reality.

This is an important point, so I will repeat it. It is a basic mistake to *first* define or describe some entity whose existence does not depend on physical things, and *then* try to derive or give reasons to accept the existence of this entity. Rather, we must take the attitude that reality is there to be discovered and responded to, and our role is principally one of discovery and response. To use a theistic term such as 'God' rather than 'liberty, equality, fraternity' or 'the matter-wave fields that quantum physics describes' is to make the claim that reality is as much or more about high-level principles such as love, justice, and mercy as it is about low-level principles such as basic physics and laws of motion. It is also to venture one step further, and affirm that what we discover is that this Real actually can be merciful and can love.

That last step, it should be admitted, is a strange step. I mean strange in the sense of subtle, hard to describe or analyse, and yet perfectly accessible to people with a sense of seeking, and perfectly ordinary as a matter of healthy and widespread human expression. I am not, in this chapter, explaining why I for one take it. I am simply saying what sort of a step it is. And I will now say a little about what forms of talking about God it suggests.

First, it suggests indirect forms—parabolic forms—Zen-like stories and hints and prods. Ways of saying, 'this is where your life and my life really springs from; do you see?'

Next, the danger of heaviness, which must be circumvented by continual efforts at lightness. Not, 'God forgives you with all the weight of ultimate authoritarian power', but, 'look: the spring bubbles up from

---

deepening recognition. After a prologue which hints at the possibility of a shared creative relationship, the *Tanakh* starts from justice as the forefront idea ('your brother's blood is crying out to me from the ground'), but this is immediately qualified by mercy and gradually extended into the intimate partnership which love involves ('what does the Lord require of you but to do justice, and to love kindness, and to walk humbly with your God?'). But don't expect all this laid out in a philosophical treatise; it is worked out in a thousand-year jumble of human folly and learning and arrogance and hope, painfully slowly giving way to a principled loveliness. The Bible is nothing if not real about human hearts.

the rock; the trees blossom afresh; the one who holds us forgives as we forgive, only more so'.

Next, forms of talking which make an effort to cross cultural boundaries, including boundaries from the past to the present, and an effort to give a sympathetic hearing to what others are saying or have said.

Next, forms which adopt a generous spirit. Think for a moment of all the efforts at doing something creative that are going on in any minute in the turning of this astonishing, rambunctious, confused, and confusing world. Somewhere a Kenyan mother is grieving for her lost son while also cooking for her living son. A million teachers are figuring out some sort of lesson plan, or juggling the needs of thirty different unquenchable children. A lively entertainer such as Lady Gaga is coming up with a way to be cool that is also a way to be forgiving. Another pop star is being helped by a humble stagehand as they struggle with self-doubt before a big gig. A billion parents sing their children to sleep. Engineers supervise building projects; landlords fill out order forms for beer. Midwives put babies skin to skin with happy, exhausted new mothers; undertakers meet their fragile clients with quiet grace. If the reality that holds all things has the sort of coherence and integrity that allows that same reality to also know these things, then what a marvellous, tender, beautiful knowledge that is.

And what an intensely painful, properly and rightly angry knowledge it must also be, when one considers all the injustice and bitter ill will also going on.

But the most lasting note must be forgiveness and the hope that goes with it. Think of an oratorio by J. S. Bach, and also an opera by Mozart: again and again, it is forgiveness that holds the key, and that opens the future. Forgiveness is the hidden theme in very many great works of literature, and it lies at the heart of the songs that do justice to our experience, and that move us. A tribute song, such as Pink Floyd's 'Shine on You Crazy Diamond', which is a tribute to the former band member Syd Barrett, dwells on whatever was positive about the songwriter's friend, skating lightly over what was frustrating about him. To talk of God is not to tack God on to the end of songs of human empathy, a tacky intrusion into another's honest grief. It is to say: 'Yes; this is where our life lies; this points us to inexhaustible hope and unquenchable forgiveness; here we recognize the face that can deserve our love.'

But the right to speak that way rests, in the end, on its truthfulness and on no other foundation.

A Principle Near the Heart of All Genuine Spirituality

Nonviolence. Non-violence. Nonviolence, nonviolence, nonviolence, nonviolence, nonviolence, nonviolence, nonviolence. Non-violence; non-violence. Nonviolence? *Nonviolence.* Nonviolence.

Nonviolence. Non-violence: non-violence; non-violence; nonviolence, nonviolence. Forthright speech but non-violence. Nonviolence.

Nonviolence.

Resistance; nonviolence. Nonviolence: non-violence—nonviolence—nonviolence. Nonviolence, nonviolence, nonviolence, nonviolence, nonviolence, nonviolence, nonviolence. Non-violence, nonviolence; non-violence. Nonviolence. Nonviolence. Nonresistance! Nonviolence . . . nonviolence . . . nonviolence.

Non-violence; nonviolence.

Nonviolence. Non-violence: non-violence; non-violence; nonviolence, nonviolence; non-violence. Nonviolence. Resistance. Resistance. Considered, controlled, moderate violence. Nonviolence. Non-violence. Nonviolence, nonviolence, *nonviolence, nonviolence,* nonviolence, nonviolence, nonviolence. Non-violence; non-violence. Nonviolence. Nonviolence. Non-violence. Nonviolence . . . nonviolence . . . non-violence. Non-violence; nonviolence. Non-violence: non-violence; non-violence; non-violence, nonviolence; non-violence. Nonviolence.

Nonviolence: non-violence. Nonviolence, nonviolence. Nonviolence, nonviolence. Nonviolence, nonviolence, nonviolence, nonviolence. Non-violence; non-violence. Nonviolence. Nonviolence. Non-violence. Nonviolence. Non-violence: non-violence; nonviolence; non-violence, nonviolence; non-violence. Nonviolence.

Nonviolence.
Non-violence.
Nonviolence.

Nonviolence.

Omnipotence

My father-in-law, when young, was
eager for butterflies and moths.
It was jolly fun to sweep to and fro
and ping such fluttering space-invaders.
He'd catch and collect what he could.

The first row he fixed in a box
with a pin marked 'omnipotent'.
The skeletons held in black and white.

Later he would use a liquid glue
called 'love' (®™).
It flowed at first, then its
volatile substances evaporated
and it set like enamel.

As an adult he had a box in the garden,
which helped him to continue his investigations.
The evening moths would gather round
and show themselves by his light.

But eventually he put that away.

When older he was caught by an
iridescent desire to sidle off.
He loved to find butterflies
amongst the wild flowers and in the summer lanes.
He would make expeditions;
his son would join him when he could.
They did not bring a net.

13

The Unframeable Picture

A Short Story

Daisy loved to draw. She drew all the time. Big pictures, small pictures, little doodles, cartoons. Every kind of drawing. Whenever mealtime was over, or school was over, or when she got home, she would rush to the paper drawer and get out lots of paper and put it on the table. Then she would grab her pens and pencils, or crayons and felt tips, or paints and paintbrushes, and get to work. If there was no clean paper, she would get scraps of paper and use those. Or if she couldn't find any scraps of paper, then she would use old cornflakes packets, or the backs of letters from school, or anything she was allowed to use, and a few things she was not allowed to use.

After a while Daisy got to be very good at drawing. And she got more and more ambitious. She began to fill great big sheets of paper with marvellous swirling lines, and intricate jagged patterns, and strange invented plants and animals all rushing about in herds, or climbing trees, or splashing in water.

One day, after Daisy had been busy drawing all afternoon, she stepped back to look at her latest picture, and she realized it was something rather special. It had a certain something that she had never seen in a picture before. It was a feeling of depth, like a hologram. It felt as if you were not so much looking *at* the picture as looking *into* the picture. When you moved your head around, the picture seemed to move in some strange way. Although Daisy knew that it couldn't really do that, nevertheless that is what it looked like.

Daisy was very struck with her picture, so she took it along to school to show her classmates. They all looked at it, and they agreed it did look interesting, but nobody was quite sure what it was a picture of. When the teacher saw it, she said it was a remarkable picture, and, if Daisy liked, then she would put it in a frame and they would hang it on the wall. Daisy had a feeling that this might not work, but she agreed to try it. So the teacher cut out a frame from card, the right size to fit around Daisy's picture, and she put it around the picture. Then she stepped back to admire the effect. But Daisy could see immediately that something had gone wrong. The picture didn't look right with a frame around it. It looked flat. It didn't look like you could go into it any more. Somehow the frame was spoiling it.

'Shall I put it on the wall?' asked the teacher. The teacher was asking this because she had promised, but in fact she could tell that the picture didn't look very good any more. Daisy looked down and shook her head.

'I don't think I would like to put it on the wall after all, please, Miss', she said.

The teacher agreed, and Daisy took her picture home again. But as soon as Daisy unrolled the picture and put it on the kitchen table, without a frame around it, the picture looked wonderful again. Daisy gazed and gazed at it. She felt almost dizzy looking down at it. She loved it. She decided she would show it to her best friend Rufus as soon as she could.

The next day, Rufus came over. He was very pleased to see the picture. He got very excited about it.

'It's amazing, Daisy!' Rufus exclaimed. 'It looks like you could walk right into it! Let's put it on the wall!' So Daisy went to get lots of bits of sellotape and Blu-Tack, and, with much difficulty, she and Rufus managed to get the picture stuck to the wall in Daisy's bedroom. It looked great! Sort of swirling and waving. Sometimes they thought it was of a tree, and sometimes they thought it was of a shoal of fish, and sometimes they thought it was stars in outer space. It looked as if there was a big hole in the wall and you could see through it, like looking through a big window.

The picture stayed there on the wall for a long time. It was there while Daisy and Rufus and their other friends played. It was there when Daisy went to sleep, and when she woke up. It was there when she went out, and when she came in. She never got bored with it.

When Daisy got a bit older, she decided she would like to show the strange picture to the professors in the big city. So she carefully took it down from the wall, and rolled it up and put it in a big bag, and went into the city on the bus.

When she got to the gate of the university, Daisy asked politely if she could show her picture, so that she could find out what it was a picture of. The porter at the gate telephoned the secretary of the chancellor of the university, and the secretary of the chancellor of the university went to ask the chancellor, and the chancellor, who was busy on the phone, pointed to a rule book. Then the secretary phoned back to tell the porter, and he told Daisy. 'You may bring your picture in, as soon as you have got it framed', he said. 'All pictures on display in the university galleries must have their own frame.'

Now Daisy could remember what had happened back at school some years ago, so she was not sure if she would be able to find a frame that would work for her picture, but she agreed to try. So she went off into the city centre to the artists' shop, and inquired if they could help her find a frame for a picture. The manager of the shop brought out various frames, and he and Daisy tried them each on the picture. But none of the frames worked. Even though they fitted neatly around the picture, as soon as the picture was framed, it did not look right any more. It lost its depth and its movement and meaning. Seeing what was happening, Daisy thanked the shopkeeper and said she would not buy a frame after all.

Daisy went back to the university and asked to speak to the chancellor. The porter said that the chancellor was too busy, but one of the professors would come and speak to her.

When the professor came to the door, Daisy began to explain. 'I have brought my picture to the university, and I hope I can show it, so that I can understand it better. But the porter will not let me bring it in until it has a frame. But I do not want to frame it. The frame spoils it.'

'But how can we look at a picture that is not framed?' replied the professor. 'All pictures in the university have to have a frame. It is a fundamental principle of our work. Only with a frame can we begin to discuss the pictures. Here, if you like, show me the picture and I will frame it for you.' A little hesitantly, Daisy unrolled the picture. The professor quickly fetched a frame and put it down over the picture. Daisy's heart sank as she saw the picture go wrong again. The professor looked at the picture within the frame he had provided.

'Oh yes, I see,' he said. 'It is rather simple, but I can see why you might have held on to it. I'm afraid it won't be possible to display it at the university, however. We need work with more depth.'

The Unframeable Picture

'Yes, yes,' said Daisy. Because when the professor mentioned depth, she wanted to show him what happened when the picture had no frame. Without a frame it had plenty of depth. 'Look, I will show you,' she said. 'We just need to take the frame off.'

'But we cannot do that,' said the professor. 'How can I explain? Pictures have to be framed. Otherwise we cannot talk about them and compare them.'

'No, please let me show you,' said Daisy. And saying this, she lifted off the frame, and once again the picture got its quality back again. Her heart leaped every time she saw it. But the professor was not looking any more. He had begun to lose interest.

'Now, if you don't mind,' he began to say, 'I have a lecture to attend to,' and he turned to leave. As he did so, the porter caught sight of the picture lying on the table, and he gave a start. 'That is a splendid picture you have there, young lady,' said the porter. 'I never saw anything like it.' Daisy nodded, a little sadly.

When Daisy got home she went round to visit Rufus, to let him know how she had got on. 'Don't be downhearted,' said Rufus. 'We can try again. I'm sure that once they actually look at the picture, they will be very interested.'

'But the rule is that all pictures have to have frames,' Daisy replied. 'And you see what happens. Our picture always changes when you put a frame around it. It spoils it.'

'I know,' said Rufus. 'Let's have another look at it anyway. It's such a wonderful picture.' Daisy got the picture out and she and Rufus unrolled it on the kitchen table. As they did this, they were standing at opposite corners of the picture, and as they reached around to hold the picture down, for a moment their two pairs of arms formed a sort of rough frame around it. 'Oh look,' cried Daisy. For when they made this frame with their arms, the picture didn't go all flat-looking. It still looked right. In fact, if anything, it looked even better.

'That's lovely,' said Rufus. And they stood for a moment gazing into the picture. They were holding hands around it. Then after a few moments, they suddenly realized this and they both gave a start, and they blushed and let go hands.

'Well, I'd better be going,' said Daisy.

'Why don't you try again tomorrow?' said Rufus.

'I might.' And Daisy went back to her home.

The next day she was busy with other things, and for a few days after that. But eventually she plucked up courage to try her luck at the university again. When she got to the gate, the porter greeted her.

'Hello Daisy, have you brought that marvellous picture along again? I'm afraid they won't look at it, you know. Not till it's got a proper frame. Can I see it while you're waiting?' So Daisy let the porter look at the picture while she waited for one of the professors to come. Eventually the professor of history came along. She was a tall lady with a rather severe manner, but she was willing to show Daisy's picture to the other professors.

'This is your last chance, mind,' said the professor of history. 'Follow me.' First the professor of history fetched a historical frame, and put it round the picture. 'Hmm,' she said. 'It is a sort of historical image, I see, but it is poor history, I'm afraid.'

Next the professor of science came. He put a scientific frame on the picture. 'Well, young lady,' he said in a bright, energetic voice, 'I do not like to disappoint you, but you must understand that this is a poor attempt at science.' And after saying that, he hurried out.

After that the professor of philosophy put a philosophical frame on the picture. 'No, no, I'm afraid that won't do,' she said. 'Your picture is some sort of philosophy, but it is poor philosophy.'

Then the professor of poetry came along with a poetic frame. 'It is bad poetry,' he announced. The professor of art said it was shallow art, and the professor of astronomy said it was outdated astronomy. But every time Daisy could see what the problem really was. The picture just did not work with a frame. You had to look at it without a frame. She tried to explain, but the professors were rather busy and eager to get on with their work, and she thought they had already been quite kind to spare some time to look at her picture. So she didn't insist. Eventually the professor of history brought her back to the gate.

'Well, I feel we have been very thorough,' said the professor of history. 'I'm sorry we cannot receive your work, but thank you for showing it to us.'

'Thank you, professor,' replied Daisy. But before she turned to go, she thought she would try one more thing. The professor of history seemed to be a kind woman, even though she had a severe manner. 'May I just show you one more thing? It won't take a moment.' The professor kindly agreed to wait, and Daisy quickly spread the picture on the ground by the gate, and asked the porter to stand at one corner. Then she held the porter's hands and she quickly made a simple frame for the picture using his arms and hers. Suddenly the picture looked more marvellous then ever. The professor of history gasped. She immediately stepped closer, and almost stepped right into the picture. Then she took one hand of Daisy's, and one hand of the porter's, and the three of them stood in a little circle gazing at the picture in the middle. The professor of history started to cry.

'Now I see, oh now I see,' she said. 'I'm so sorry I could not see it before.' Daisy didn't mind. She was happy.

After that, each day Daisy would go into the city, and instead of going to the university, she went to the square in the middle of the city, and showed her picture to anyone who came along. Soon lots of people began to look forward to her visits. Daisy would put the picture on the ground, and all the people who wanted to look at it would gather round and make circles and squares and other shapes around it. Some would stand in intense concentration, others would laugh and dance. Eventually they made a big conga line. Daisy looked on and said to Rufus, 'Look! the shapes we are making are like the picture, don't you think?'

14

A Farewell to Hume

> A well-known argument from David Hume is presented and refuted. This is the idea that the natural world may be self-contained, for all we know, and theistic religious claims cannot help but be superfluous. This is also advocated by Richard Dawkins. I explain that these arguments fail owing to what amounts to a false premise. It is subtle, however, because the false premise is in the very way the discussion is framed. If one assumes that when we are talking about God we are talking about abstract intellectual tools, then one goes wrong. I invoke various witnesses to show that thoughtful religious language operates differently.

In this chapter I consider a well-known argument that was briefly presented by Thomas Aquinas, and later elaborated by David Hume, and a version from Richard Dawkins which amounts to the same argument again. The argument concerns the failure of religious discourse, especially the notion of Creation, or Creator, on the grounds that it is unwarranted and superfluous. Thomas Aquinas had already given a response many centuries before Hume presented his case. I think that Aquinas's work went a long way to answering it, but that more needs to be said. I will argue that this argument can be answered, but only by admitting the failure of all passive language to capture the whole of the reality with which human life is concerned. In short, Hume was right to say, as R. S. Thomas phrased it, 'The mind's tools had / no power convincingly to put him / together', but this does not mean the life of the whole person has no power convincingly to encounter and respond to God.

The Scottish philosopher, historian, and diplomat David Hume (1711–1776) might fairly be called one of the 'giants' of philosophy, and his writing on religious questions is widely regarded as both important and convincing—more so, for example, than that of Bertrand Russell.

Science and Humanity. Andrew Steane, Oxford University Press (2018).
© Andrew Steane. DOI: 10.1093/oso/9780198824589.001.0001

In this chapter I am not going to assess Hume's contribution in the round, nor ask where he stood in his own personal commitments. Rather, I will take a look at one particular line of argument that is generally felt by atheists to be cogent and convincing. This is the argument that says, broadly speaking, 'Why go further?—the natural world of physical things and events is evident to the senses and there is nothing to be gained by positing anything else.' I will let Hume make the point in his own words in a moment. I have called this chapter 'a farewell to Hume' because I think one can see his point and accept what he is trying to say, and yet move on without taking leave of a proper theism. Also, I would wish well to David Hume, I would wish that he fares well; I say this as a way of acknowledging that he probably felt unable to speak as openly as he would wish, in his own lifetime, owing to religious intolerance, and what he did say needed to be said, fairly and openly.

14.1 Introduction

Religious communities may fall into intolerance, but they don't always, and in fact theologians had been discussing the point aired by Hume quite openly for many centuries before he was born. That is to say, essentially the same argument had already been made by the thirteenth century, and St Thomas Aquinas (*c.*1225–1274) addressed it in his *Summa Theologiae*. To be clear, Aquinas considers the issue as an objection which he seeks to answer, whereas Hume asserts (or strongly implies) that it has not been answered satisfactorily.

David Hume presented the argument in a philosophical discussion, and more recently it was emphasized again by Richard Dawkins. I will quote from Hume in the first instance since he presents the point at greater length than did Thomas Aquinas, and he claimed that it had not been well answered at his time of writing (though this is disputed). Hume's writings concerning problems of religion 'are among the most important and influential contributions on this topic', states the Stanford Encyclopaedia of Philosophy,[46] and his argument against the value of religious thinking captures one of the most widely felt and important problems in this area. Dawkins has taken Hume's philosophical approach and recast it in a more scientific style; this results in essentially the same line of reasoning but in a form that is more concrete and easier to summarize for scientifically minded people.

In this argument, and others like it, the word 'God' is used, or else a brief descriptive phrase is used, such as 'Author of Nature', or 'Divine Being'. The chief cause of confusion in this area is, I think, the lack of a deep grasp of what the word 'God' means, or can mean, in careful religious discourse. Briefly, if one defines some sort of entity for our inspection and calls that entity 'God,' then one ends up concluding that the entity is unreal or unworthy of the name (like the wooden idols scoffed at by Isaiah). But if one realizes that the reality we are called to recognize as God is not engaged with that way, then one is left free to engage without getting distracted by bad theology. This brief summary is what the present essay elaborates.

One may determine what anyone means by a word by paying attention to the way they use that word. I hesitate to quote Dawkins on this subject, because his usage is almost entirely drawn from unreflective religion. He makes, I think, little attempt to grapple with the subtlety of serious theological discourse. However, since he has expressed an argument that many find compelling, it merits a response. Hume's writing is more careful but in some respects is similarly flawed, as I will discuss.

A reply to Dawkins's argument about God being highly improbable can admit that 'God' as conceived by Dawkins is highly improbable, but the more interesting question is, does his argument also apply to a more thoughtful meaning of the word 'God', and is that more thoughtful meaning the one advocated in the *gospels* and experienced by various constructive and creative communities of people over a long period of time? I will argue that it does not, and it is, respectively. That is, the argument of Hume and Dawkins does not apply to God, as God is understood in Christian response (properly so called).

Section 14.2 presents the argument that I wish to address, first as articulated by Hume, then in a form I give in my own words and which I think captures the point advocated by Dawkins. Section 14.3 briefly presents the point considered by Thomas Aquinas, and a response from one contemporary philosopher. Section 14.4 presents the core of my response, which is similar but which is presented at greater length. My main point is that the very way a question is framed can itself be misconceived, and this is going on in the arguments in this area. The language of persons and their relationships is fundamentally different from the language of scientific models or philosophical abstractions; it is the failure to allow for this that leads to nonsense, not the failure of personal language *per se*. In section 14.5 I illustrate and support my case by examining the way that language about God functions in the experience of a series of witnesses. Section 14.6 and Chapter 15 then discuss further what such language is and is not about.

14.2 The Argument from Lack of Explanatory Power

Let us begin with Hume. In his *Dialogues Concerning Natural Religion*[35] the character Philo addresses the idea that the natural world has its origin in another and non-contingent Being:

> How, therefore, shall we satisfy ourselves concerning the cause of that Being whom you suppose the Author of Nature, or, according to your system of Anthropomorphism, the ideal world, into which you trace the material? Have we not the same reason to trace that ideal world into another ideal world, or new intelligent principle? But if we stop, and go no further; why go so far? why not stop at the material world ... If the material world rests upon a similar ideal world, this ideal world must rest upon some other; and so on, without end. It were better, therefore, never to look beyond the present material world. By supposing it to contain the principle of its order within itself, we really assert it to be God; and the sooner we arrive at that Divine Being, so much the better. When you go one step beyond the mundane system, you only excite an inquisitive humour which it is impossible ever to satisfy.
>
> To say, that the different ideas which compose the reason of the Supreme Being, fall into order of themselves, and by their own nature, is really to talk without any precise meaning. If it has a meaning, I would fain know, why it is not as good sense to say, that the parts of the material world fall into order of themselves and by their own nature. Can the one opinion be intelligible, while the other is not so?
>
> ...
>
> The first step which we make leads us on for ever. It were, therefore, wise in us to limit all our inquiries to the present world, without looking further. No satisfaction can ever be attained by these speculations, which so far exceed the narrow bounds of human understanding.

Just before the above extracts, Philo states:

> If Reason (I mean abstract reason, derived from inquiries a priori) be not alike mute with regard to all questions concerning cause and effect, this

sentence at least it will venture to pronounce, That a mental world, or universe of ideas, requires a cause as much, as does a material world, or universe of objects; and, if similar in its arrangement, must require a similar cause.

Although Hume was writing in the eighteenth century, it should not be assumed that his argument is limited to addressing the kinds of ideas that were then prevalent. He has expressed what many people now feel is a perfectly sensible attitude. The idea that the 'parts of the natural world fall into order of themselves and by their own nature' is not something we can know for sure, but for many purposes it is sufficient as a working assumption. It enables us to get on with our lives, and it is, arguably, a sufficient basis for science (but see[52]). Furthermore, the feeling of 'things falling into place' expresses very well what we experience when we gain scientific insight.

Dawkins's contribution is to reassert the argument in fresh language, making clear its force, and insisting that it be grappled with and not ignored:[23] 'Any Designer capable of constructing the dazzling array of living things would have to be intelligent and complicated beyond all imagining. And complicated is just another word for improbable—and therefore demanding of explanation.' In short, such a Creator would be 'even more in need of an explanation than the object he is alleged to have created'.

Dawkins writes in a somewhat journalistic style that does not always trouble to get precision. His own stated view is that in fact no intelligence would be required to make the universe as it is, which contradicts his statement above. However, if one thinks that the deep patterns of the natural world do not explain themselves—including, for example, those that make Darwinian evolution possible—then one is left with something (the existence and structure of the natural world) which does, at the least, invite reflection on why it might be so, and religious language tries to contribute.

There is debate amongst biographers about the degree to which Hume's character Philo expresses Hume's own overall position, or whether Philo serves to allow Hume to make some pertinent points while keeping his own convictions quiet, so that they don't intrude on a philosophical argument. Therefore one should not simply identify Hume with Philo. Nevertheless, the argument put into the mouth of Philo is an argument Hume is making, so I will refer to it as his argument.

Dawkins's point is not precisely the same as Hume's, but they are closely related. Hume argues that if anything 'contains the principle of its order within itself', then that thing may as well be the natural world, for all we know, and there is no insight to be gained by going beyond this 'mundane system'. Dawkins argues that the 'insight' offered by an appeal to a 'Designer' is no insight because it attempts to answer a question by introducing concepts just as much or more in need of explanation than was the thing originally asked about. This is the same point as Hume's in all ways that matter for the purposes of the present essay. Both consider that an 'Author of Nature' would be as much in need of a cause or an explanation as the natural world. The claim is not that one has an alternative explanation for the existence and nature of the physical universe; it is rather a claim about what kinds of explanation might in principle warrant a hearing. The claim is that rather than theist talk, of whatever form, it makes better sense to say that qualities such as intention and personhood are products of complex systems that emerge from wholly impersonal and mindless origins.

Like many people, I have puzzled over this argument. On one hand, one can see its force. But on the other, it seems somehow to veer off to the side of well-developed theological discourse. It adopts a way of speaking that does not get much purchase on the hard-to-describe sense of connection to God that is at the heart of what we mean by religious faith.

A clarifying statement has been made more recently by Prof. Dawkins, when he states clearly that he regards religion as a form of science. He considers that religion is an attempt to provide explanations, but it is fundamentally misconceived because its explanations are no explanations at all. They are not simplifying; they merely replace one not-understood thing with another even-less-understood thing. Such attempts to understand introduce much superfluous baggage [12] and, all told, make matters worse. This further statement is helpful, because it correctly identifies what the real issue is here. The issue is, if religious language is not a form of bad science (as atheists assert), then what is it? If it makes sense, what sort of sense does it make? Because, on the face of it (say Dawkins, Dennett et al.), it makes no sense at all.

The form of reaction that the philosophy of religion can take was stated eloquently and correctly by Phillips:[50] 'But once having shown, as Hume showed conclusively on so many issues, what religious activities do not mean, the next step, surely, is to ask what they do mean, not to doubt whether they can mean anything at all.'

In Chapter 1 I stated that religion is an effort to recognize correctly what we truly depend on, and to seek what can properly be aspired to, and to respond appropriately to what we find. The activities we call religious do not have as their main aim the elucidation of the mechanism that allows phenomena in the natural world to take the forms they do. It is not that one is not interested in such mechanism; it is simply that that is not the only thing one is interested in. Furthermore, one realizes that our wider context is not merely a playground for the satisfaction of our curiosity. The world is not our servant. Religious activities are intended to help people negotiate other matters, especially to do with what their own life is to be directed towards and where its meaning chiefly lies. When it is done well, this involves pressing against the boundaries of language itself. This is not a fruitless exercise, and it is not wholly different from other areas of academic discourse, such as ethics, aesthetics, and jurisprudence. But it is not science. It should not be mistaken for science, and nor should this be regarded as a defect.

Before we address this, it will be helpful to have the argument advanced by Dawkins and others stated in modern language. For clarity, I call it 'the argument from lack of explanatory power'.

Theorem 1: The argument from lack of explanatory power (LEP)
Broadly speaking, simple things can be regarded as fundamental, probable, and not in need of explanation, whereas complex things should be regarded as not fundamental, and either improbable or in need of explanation in terms of simpler things. God falls into the class of complex things, and hence is either improbable or is in need of explanation in terms of simpler things. But if the latter, then God does not merit the name God in any meaningful sense. It follows that the former holds: God is improbable. Sufficiently so, that no rational person need concern herself or himself with the existence of God, since it is a premise whose validity is so improbable that it serves no useful role in intelligent discourse, and consequently it is untenable.

I have presented the argument here in my own words. In *The God Delusion**) Dawkins gives great weight to this point about explanatory power and simplicity, stating:

> The argument from improbability—the 'Ultimate 747' gambit—is a very serious argument against the existence of God, and one to which I have

---

* I quote from this book merely in order to respond to an argument in it that merits a response; this should not be taken to imply respect for the book as a whole, which I think misrepresents history, faith, and scholarship in seriously objectionable ways.[20]

yet to hear a theologian give a convincing answer despite numerous opportunities and invitations to do so. . . . If the argument of this chapter is accepted, the factual premise of religion—the God Hypothesis—is untenable. God almost certainly does not exist.

Daniel Dennett also refers to the argument in enthusiastic terms:[25] 'Dawkins' retort to the theorist who would call on God to jump-start the evolution process is an unrebuttable refutation, as devastating today as when Philo used it to trounce Cleanthes in Hume's Dialogues two centuries earlier.' Note that this comment from Dennett is, strictly speaking, aimed at one gambit in the context of biology. However, from his wider writing, Dennett is in broad agreement with Dawkins on the importance and power of this type of reasoning when it is used more generally to refute theism. Plantinga, on the other hand, is not so impressed:[51] 'I am sorry to say that it doesn't seem to me to be a masterstroke at all. Dawkins' retort is neither unrebuttable, nor devastating, nor even relevant; it irrelevantly addresses a claim not at issue.' As I understand it, what Plantinga (a professional philosopher) means here is that Dawkins's argument succeeds in showing that various claims about the origins of complexity in the world do not hold water, but no one in academic life and familiar with the territory is making those claims. It is true that you can indeed find dubious claims (about macro-evolution, for example, and about the origins of life) in many works of popular Christian apologetics, and perhaps the reader may feel that this is Dawkins's real target. I think that Dawkins is trying to make a stronger point, one that will allow us to wash our hands of the whole business of God. But before addressing this, for clarity let me say that I welcome the demolition of dubious arguments of whatever kind.

Some people think that macro-evolution, or the beginnings of life, cannot have happened through the sorts of natural mindless process that we find elsewhere in biological history. I think that is wrong. No one knows for sure, at the moment, but I think probably life got started because the universe was pregnant with the possibility of it and ready with the wherewithal to support it. The fact that the natural world has such propensities remains a powerful indicator that it is the outcome, or the expression, of an extraordinarily creative origin or root. This also heightens, rather than diminishes, the sense in which God may have regard for the world, since it makes the world all the more expressive and valuable. So, OK, let's move on; this aspect of the natural world is not in dispute (with me at any rate); I am interested in other matters.

Some of the responses to LEP have been helpfully summarized by Richmond,[55] and he adds his own critical remarks. I agree with him that various of the responses he lists remain unsatisfactory, and I have some, limited, sympathy with Dawkins's claim that he too has not heard a convincing response to LEP. Limited, because this issue was already addressed in the thirteenth century by Thomas Aquinas, and also Hume's work has been widely recognized and there can be hardly any theologian or philosopher since the eighteenth century who has not thought about this, and come to some sort of accommodation with it. However, the argument is clearly not resolved satisfactorily for many people, and it may be helpful to present a response in fresh language.

I don't agree with all Plantinga writes in 'Darwin, Mind and Meaning', but he does provide what I think are pertinent pointers to what are the issues here, and they are not the ones that Dawkins seems to assume. The problem here occurs right at the outset. Dawkins thinks that when we speak of God, we are trying to construct a simple-but-adequate, abstract, and passive* account of the patterns to be found in the physical world and in mental abstractions. In other words, he thinks we are doing science (and he implies that this sort of activity is the best way we can spend our time). He assumes that this is the type of intellectual activity going on in religious discourse, and some sort of Supreme Being or Eternal Mind is introduced as an explanatory hypothesis. He then points out that such a hypothesis is totally unconvincing, because it is at best circular and at worse just makes matters less well understood.

The reply is: well of course. The point is that you don't have to be very astute to see this particular point, and so of course we have all seen it. This is the essence of what Plantinga means when he says that LEP 'irrelevantly addresses a claim not at issue.' The mistake Dawkins accuses religion of making is that it sets out to explain organized complexity, but in fact just assumes or postulates it. Plantinga replies that this is such an obvious mistake that it is hard to credit that the long list of gifted intellectuals who embraced Christianity (for example) never noticed it. We might be stupid, but we are not that stupid.

That reply is not yet an argument, it is merely a riposte, but I have included it in order to invite the reader to take an initial step. This initial step is one of receptivity: it is to be receptive to the idea that maybe religious thinking is not ignorant. Christian theology has not

---

* By 'passive' here I mean an account that we can put forward and peruse while holding ourselves passive, one step removed from the account, because the account is an abstract thing that we can sit in judgement on.

been ignorantly unaware of Hume's argument and its antecedents, and has not ignorantly chosen to ignore them. Many so-called apologists do indeed write very ignorantly in this area, but not all. However, the point is a not a trivial one, and the reply is not quick or easy. The reply that I think theology can make requires some care in its statement. It invokes ways of putting things together, of making sense, that are not precisely analytical ways.

14.3 An Example Witness

Peter van Inwagen[74] presents philosophical and religious reflections on the argument considered as a possible objection to theism by Thomas Aquinas in Part 1, question 2, article 3, objection 2 of his *Summa Theologiae* (1274).[8, 38] This objection runs:

> Objection 2. It is, moreover, superfluous to suppose that what can be accounted for by a few principles has been produced by many. But it seems that everything we see in the world can be accounted for by other principles, without supposing God to exist. For all natural things can be accounted for by one principle, which is nature; and all voluntary things can be accounted for by one principle, which is human reason or will. Hence, there is no need to suppose that a God exists.

Aquinas brings this in as an objection to which he has a reply. Let us note that this 'Objection 2' is close enough to Hume's point as to amount to the same objection, and we have already stated that the differences between Dawkins's and Hume's arguments are not pertinent to the present essay. Van Inwagen refers to this objection as the superfluity argument; for present purposes this and LEP can be regarded as synonyms.

In order to understand Aquinas's reply to the superfluity argument one must invest time in acquiring a set of ideas which he lays out in his work. These all surround and fill out the notion that God is not like things; God is not a further thing in addition to all the other things, not a sort of extra-powerful thinking entity residing somehow alongside the rest of what is real. I will return to this in the next section.

Van Inwagen first carefully considers what would be a justified response, supposing that the superfluity argument is accepted. He then goes on to consider the fact that science, and the scientific method, is not obviously able to give an account which provides underlying causes for physical existence itself. This may be because the phrase 'underlying cause for physical existence itself' is a meaningless phrase, but it is not obviously meaningless. An atheist will probably take the view that every observed fact has either a purely natural explanation or else no explanation at all. (Note, this is not an argument for atheism; it is the premise of atheism.) Accordingly, van Inwagen turns next to ask: 'should we accept the principle that everything we observe either has no explanation whatever or else has a purely natural explanation?'.

Now, in order to address this question, van Inwagen begins to do something that is at the heart of the point I wish to make in the present essay. He admits that he does not believe in God because he, Peter van Inwagen, made a hypothesis and then assessed the merits of that hypothesis; he arrived at his belief another way. This does not mean that this other way (to be described) is irrational, because he also has plenty of other beliefs which he arrived at similarly and which people do not normally consider to be irrational. This includes the belief that there are material things and not just minds (à la Berkeley); that other people have an inner mental life like his own; that men and women are intellectual equals; and he gives further examples.* He believes these things, he says, because they came to him gradually by ways that he is unable to trace in any detail, and although he can give rational arguments to support all these beliefs, he senses that the rational arguments don't do justice to the way the beliefs actually function in his experience. And his belief in God is like that.

14.4 Resolution: The Full Expression of Human Personhood

14.4.1 Words, Usage, Categories

I now turn to LEP. I would like to make it absolutely clear at the outset that I do not consider that there is an entity called 'God' that might or might not exist, and we have to scurry around gathering evidence one way or the other. I consider that that way of speaking is confused. It is confused because it has to begin by giving some sort of definition of the word 'God', so that one has some notion of what it is whose existence

---

* The same point is made by Plantinga:[51] 'There are plenty of other things we rationally accept without argument—that there has been a past, for example, or that there are other people, or an external world, or that our cognitive faculties are reasonably reliable. Moreover, one lesson to be learned from the history of modern philosophy from Descartes to Hume and Reid is that there probably *aren't* any good arguments for these things—but we are still perfectly rational in accepting them.'

is in question, but the definition provided is not one which matches to anything that deserves the name 'God'. If the thing being called 'God' is so paltry that it might not even exist then it is utterly undeserving of the title. So the starting point is to reposition our outlook. The starting point is to declare that what we mean by *God* is that which is most profoundly real and objectively deserving of our allegiance, whatever that may be. In this important repositioning, we take on a different role. Instead of trying to establish the existence of something we defined, we instead look outward, seeking whatever is real and worthwhile. Our role is to discover what that may be, and to be receptive to what that may be. As I put it in the closing section of Chapter 12, *our role is not to derive the existence of reality but to discover and respond appropriately to the nature of reality.*

The phrase 'God exists' is only of limited help, because it is liable to lead and mislead at the same time. The problem with it is that it appears to suggest that 'God' is an item that might not have existed, but does. Such a way of thinking has already misconstrued the nature of the territory.

The phrase 'God exists' might be used more positively, as a way of saying 'since looking to God I have got on top of my addiction and become a better parent and friend.' In this case it is a way of saying, briefly, something rather valuable. But more often the phrase is used in the wrong way, for example when someone uses it to say or think, 'that which I now comprehend and name as "God", exists'. The phrase is used in the right way when someone uses it to say or think, 'the value and truth which we are seeking is before us but I am on the way; I am confident that there is more to be discovered, and this is worth reaching for.' To be precise, language about God means something profoundly yet subtly transformed from that. The version I just gave could be uttered by an atheist. What theism does is to move from, 'the value and truth which we are seeking . . . ', to begin to say, 'the relationship of trust which we are seeking . . . '. I call this 'profoundly yet subtly transformed' because it moves from oneself being the judge, wholly in charge of one's own identity, to allowing God to be judge, shaping one's identity.

But how does anyone ever move from one attitude to the other? The answer is that each person does it when they choose to do it, and insofar as they are able to do it. They are able to do it when they can see that it is an honourable and a creative thing to do, or when they can see simply that it is their duty, or when they are carried along by a sense of a relationship starting whether they meant it to or not.

Now let us begin to address the task of this chapter more directly.

One does not assess whether the French language is capable of eloquence by trying to write poems in franglais. Still less does one do it by shouting at foreigners in loud English. Similarly, one must not expect theism necessarily to accept all the terms which atheism may try to impose on it. A reply to Hume, or to LEP, does not take the form of providing evidence for the existence of something. It is not a sort of 'will he, won't he?' kiss-and-tell story in which some deity figure appears and disappears in a sequence of philosophical moves. A reply takes the form of providing evidence for the fact that theism is intellectually coherent. The task is to show that adopting personal language in describing our relation to what is most profoundly real does not involve any failure towards our rational duties.

In a reply to LEP one does not start out from some cautious premise about God, in which 'God' is utterly impersonal, and then try to build from there. Rather, one explains what is the nature of the case when that which gives rise to the universe of embodied truth—God—is able to meet with persons on a personal level. Let me reiterate: by recognizing such personal qualities in the truth that forms all things, one does not need to imagine that there is a 'god', a 'deity' that is like a superhuman. Rather, it is the point I made at the end of Chapter 10 when I mentioned the experience of 'intimacy and shared endeavour'. Neither should one see this claim as a bizarre jump away from the scientific story. On the contrary, it is a natural extension of the structure of ideas that we discussed in Chapters 3 to 6. The natural world does not create from nothing, but embodies from truth. That is to say, the things that come to be in the world, that have a chance of becoming physically expressed, and of expressing themselves, are not arbitrary agglomerates fluctuating into existence out of a random soup. Rather, they are imperfect expressions of wonderful truths; truths that transcend any given embodiment. That such truths can be personal, not just impersonal, is a natural extension to the patterns embodied in impersonal things, but it is an intensely challenging one.

In the following my aim is to show that the objection raised by LEP is misconceived. I will do this by explaining what is the nature of the case when personal language is appropriate. Personal characteristics cannot be derived, only recognized and borne witness to. What we can do is learn what kinds of response are appropriate when that which concerns personhood is before us. The nature of the logic here is not, 'this is so, and therefore theism is right.' The logic is, 'it is possible that theism is misconceived and wrong, but theism is at least capable of being thoroughly intellectually and morally and emotionally coherent, and therefore it is also possible that theism is right.'

I refute LEP by denying one of its premises. I do not accept the premise God falls into the class of things, where I mean that God does not fall into the class of things that can be addressed in the way adopted in LEP. I would accept that 'God' as conceived of by many people, Prof. Dawkins among them, would fall into the class of things (complex things, to be specific) that you can conjure with in such a style of argument, but God, properly so named, does not. But one must immediately add that this is subtle, because I am not here making the elementary mistake of assuming that the 'things' named in LEP have to be physical things or even supernatural things. I accept that the word 'things', in the context of LEP, is intended to be as general as it could be. But nevertheless I claim that it does not capture all possibilities because it cannot capture those aspects of reality that are inaccessible by this type of approach. As Stephen Jay Gould put it in another context,[31] 'As with so many persistent puzzles, the resolution does not lie in more research within an established framework but rather in identifying the framework itself as a flawed view . . . '. This was a comment on puzzles in evolutionary biology, but the same principle applies here. I maintain that there are aspects of what is real—there are realities—that do not fall into any class that one can adequately access or speak of in the form of language that LEP adopts. And God is such a reality.

The form of language that LEP adopts is a form commonly adopted in science and philosophy. One presents a concept, a hypothesis perhaps, and one gives arguments as to why the hypothesis is supported by other evidence, or runs contrary to other evidence. This is of course a perfectly reasonable thing to do in many areas of human life. But the point is, it is not the right approach in all areas.

To show this, I will present two more everyday examples, and then consider at greater length the example with which we are concerned.

For our first everyday example, consider how people become friends. This generally happens when people are thrown together by circumstances and find that they have interests in common. They are each open to the possibility of friendship, and intimacy begins to grow. The question I would like to consider is, is it possible to base a human friendship on a logical argument? Do we do a calculation along the lines of, 'if I am a friend to Emma, then good will come to me, therefore I will be a friend to Emma'? The problem with this is that if my supposed friendship is based on that kind of calculation, then it is not friendship because I am not really interested in Emma as a person. A true friend is not primarily seeking this kind of payback. So let's try another approach. Suppose we try the calculation, 'if I am a friend to Emma, then good will come to her, therefore I will be a friend to Emma'. But anyone who likes that calculation is already a friend to Emma (because they think that good coming to Emma is a desirable outcome).

So how do we ever get to be friends with anyone? The answer is, by not being quite so calculating, and allowing other sorts of issue to influence us. We do it by paying attention to each other, and allowing parts of our identity to become bound up with another. We give up a little of our self-determination. At no point does either partner require a syllogism in order to convince themselves that it would be rational to promote the friendship. In fact that kind of attitude risks damaging the friendship. It risks damaging the friendship because it amounts to an attempt to establish friendship on something other than a personal basis. Person B does not want to hear that person A befriended him after making a calculation that this would be a rational thing to do; he would prefer that person A's friendship is a direct expression of person A herself. Friendship blooms when each acts in and of themselves, and risks themselves. The risk is not an optional extra.

For our second example, consider the case of Shakespeare's King Lear asking his children, 'Which of you shall we say doth love us most?'. Lear is telling his children that they must each present arguments to establish a claim to love him more than the other siblings love him. What do we imagine going through Lear's daughter Cordelia's mind, faced with such a request? Shakespeare tells us her anxiety and her attitude: 'What shall Cordelia speak? Love, and be silent.' Cordelia does the calculation and rapidly comes to the conclusion that her father has asked for what cannot be expressed the way he has demanded that it be expressed. In this example, it is the very way that Lear has framed the discussion that has already undermined it. Something similar is going on in arguments about the explanatory power of God.

The difference between science and thoughtful religious discourse is similar to the difference between theorem-proving and learning to trust. When a word such as 'God' or a descriptive label such as 'Designer' is introduced in the first kind of activity, it functions, and can only function, in the way that theorem-proving allows it to function. It signifies a tool. When the same word or label is used in the second kind of activity, the activity itself implies that the word and the label is functioning differently. Now it signifies not a tool but a participant in a relationship of mutual trust (or distrust). This is a sufficiently large difference that it amounts to the same word being used for two fundamentally different types of thing. And the point is that this difference in signification is owing not to a stack of careful definitions, but to the attitude adopted by the one speaking.

When reading Hume's *Dialogues* one can grasp the broad fairness of the points he makes about what constitutes adequate and inadequate explanation. Considered as an argument about how intellectual tools are used, it is a valid argument. But in view of the above one may legitimately ask whether or not Hume has so framed his discussion that it cannot access the meanings of the phrase 'Author of Nature' that he is seeking to question—the meanings that would allow God to be an appropriate focus of personal trust. Similarly, my experience of reading Dawkins, and Dennett, and others, is a continual impression that the argument veers off to one side. It engages with various questions that those authors want to ask, in the terms they employ, but it fails to grasp or address the type of rational activity which we are engaging in when we speak of faith in God. It is as if a football player came to join a cricket match, and complained that the ball was too small, did not bounce, and hurt his foot. It is like talking to someone who is looking over your shoulder. These comments are not intended to belittle the honest questions which these and many other people raise; they are intended to help the reader understand that the issue is not resolved by a direct answer to the questions, but by showing that the questions have adopted a fundamentally misconceived attitude, like the question posed by King Lear.

Dawkins assumes that to talk about God is first and foremost to talk about an entity that exerts influences much in the same way as other entities do (electric fields, tennis balls, humans, galaxies, DNA molecules, etc.). This sounds odd to a religious person, but it is intended to be straightforward, frank, and honest. 'Is X someone or something that influences physical phenomena, or started them a long time ago?' he asks, in effect. 'If no, then I do not need to bother about X because everything that happens to me or that I become aware of happens through physical events. If yes, then scientific reasoning can be applied to X, because scientific reasoning can be applied to all phenomena in the realm of physical phenomena. So let's get to work', the train of thought continues. 'We do some thinking about the case "X = (omnipresent eternal entity capable of creative action, empathy and response)", and arrive at LEP. So we can conclude that we don't need to concern ourselves further about God.' It all sounds perfectly clear and sensible. What could possibly be wrong? Is the rest of the present volume really just obfuscation and elaborate footwork to sidestep the requirements of plain and honest truth?

Let me have an initial stab at this. 'Omnipresent, eternal'? Ye——es. I give a qualified 'yes', because asking about spatial and temporal location here is a bit like asking about the spatial location of a rainbow: anyone asking about that has not got quite the right way of understanding the situation. But within the limits of everyday language, these adjectives capture what is experienced. 'Capable of creative action, empathy, and response'? Yes, but not in the 'superman' style, not zooming in to fix problems, but rather the convener of the meetings we are in fact having, sharing the burden, working towards a more just future, and yet as stymied by blank unwillingess as we are. But, 'entity'? No. That doesn't seem to work; that word doesn't help; that's the sort of idol-making which we have learned to avoid.

14.4.2 Thomas Aquinas and Divine Simplicity

One way to avoid the conclusion of LEP is to argue that explanations in terms of God are cogent because that which is correctly named *God* has a degree of coherence that does not require or invite further analysis. This is at the heart of what Thomas Aquinas expounded in his *Summa Theologiae* (part 1, question 3), and it constitutes his 'answer to Hume'. The idea is that when we trace the processes and complexity of the world back to their source, that source is precisely not like them, and this fact extends right into the very concepts that apply. One cannot, says Aquinas, speak of God as an entity that can be said to be in any set or genus or class, because 'on the contrary, in the mind, genus is prior to what it contains. But nothing is prior to God either really or mentally. Therefore God is not in any genus.'[9] This idea takes some getting used to; it is not easy to assimilate.

Aquinas argues further that it does not make sense to talk of God as composed of parts, nor as if God could fail to exist. To God it belongs that God's very nature is to be. That is right.

Finally, bringing together these and other considerations, Aquinas argues that God is not complex, but rather indeed supremely *not complex*. This has become known, in theological technical terminology, as the idea of *divine simplicity*. It functions as an aid to exploring, or thinking through, the sheer directness with which the verb *to be* applies to God. In what sense is the word 'simple' appropriate here, though? After all, this claim is put forward along with standard Christian ideas about the very great richness of God's creative skill and the depth of God's compassion and the incorruptibility of God's justice. Can all this simply *be*, without recourse to simpler substructure?

This is close to the heart of the issue at hand; it is the point that Hume's character Philo presses home. My own reaction has been to try to get the flavour of the idea called divine simplicity, and only weigh it after doing that. This is not done quickly, however. For me it has involved wide-ranging thought over a long period. My tentative conclusion is that the flavour of this idea is like the way works of art, especially music, can be complex in one sense, and yet also attain a degree of integrity and coherence which expresses something profoundly *just so*, a lucid quality which does not require to be subdivided or analysed. Of course with our violins and oboes, cellos and clarinets, we express such musical profundities through the medium of many things woven together in a complex way, but the musical idea itself is an undivided whole, a word in a language which we can only speak when we join our efforts together, but which would be simply spoken by one whose language it was. It is very striking that something similar goes on in fundamental physics. The equations of quantum field theory would be extremely awkward to write down, and almost impossible to work with, if one did not adopt an appropriate and deeply informed set of concepts with which to express them. One needs differential calculus, tensor and spinor calculus, and the notion of operator as opposed to number, for example. Only expert readers will know in detail what I am referring to here, but the illustrative point is accessible without that expert knowledge. The point is that one finds oneself working with deep mathematical concepts, yet these highly sophisticated concepts, concepts that most people do not have the intellectual gifts to ever learn, are nevertheless felt by practitioners to have a certain elegance and just-so-ness. Like the musical example, there is here a simplicity that appears highly complex when expressed in any language ill-suited to its expression, but which is simple, in a profound sense, nonetheless.

Another example occurs in pure maths. Deep mathematical insights usually involve realizing how symbols and ideas can be brought together in an apt and telling way, and, even though to the rest of us the insight is unavailable because the tools and techniques are beyond us, to the mathematician having the insight, a beautiful just-so-ness is on view; a sense that the landscape of ideas *must* be like this. Some such combination of profundity and integrity is what, I think, the theological discussion of 'divine simplicity' is trying to assert. It has been thought through carefully over a long period, taking its place in a body of ideas called *classical theism*, which is a mainstream part of Christian theology (though not the only view). The musical and mathematical illustrations that I have used should not be construed as a statement that God is music or mathematics; their purpose is to convey the flavour of an idea about the integrity of God.

I will not present the doctrine of divine simplicity further here, because I will not rely on it, and I have some reservations about it. I consider, though, that it is part of a body of intellectual work which is intensely disciplined and honest. This body of work says 'no, no, and no again' to the anthropomorphisms of the fancy, to all the incoherent images and ways of speaking that we seem helplessly to fall into unless we exercise the utmost care when talking about God. The task it addresses is similar to the mathematical task of saying why it is wrong to think of infinity as just like other numbers, only bigger. My reservations are to do with the danger of sophistry, and the danger of adopting an inappropriate stance when we approach the task of speaking about God. I think, though, that classical theism is perfectly correct when it replies to Philo (of Hume's discourse) that to appeal to God is *not* to appeal to 'another world' in the way Philo suggests, and hence invite an infinite regress. It seems to me that Hume (through his character Philo) was indeed wrong at this point. In fact the appeal to God is an appeal to One whom we can reasonably understand to be different from the natural world in this very respect.

The idea of divine simplicity has been set out at greater length by various recent authors (see, for example,[60, 67, 27, 29]). It has to be seriously engaged with by anyone who is genuinely interested in education in this area. It informs the background of the current essay, but what I want to argue here is that this aspect is not the heart of the matter. The central issue is prior to this. The false step in LEP is not in assigning God to the wrong class (complex things as opposed to simple things), but in assigning God to any class at all. You just can't do that with God.

14.4.3 Opening the Self, Not Just the Mind

One can assign many things to sets or classes, and one can probably assign to one set or another everything that human brains are capable of fully grasping, but it does not follow from this that all that is real can be captured in that type of approach. When one assigns an individual person to a set, for example, as in 'Eve is in the set of my friends', one immediately senses that something isn't quite right; Eve is not an object at my disposal; she may be a friend but that doesn't put her in a set; she immediately jumps out and I imagine her saying, 'you can't get me pinned down quite so easily, my friend.' Obviously we use names and labels to help keep track of one another, and this is not a wrong thing to do, but the inadequacy of dealing with people in abstract terms alerts us to the type of inadequacy that can arise in attempts at abstraction. It may be that there is that which is real but which cannot be assigned by us to a set or class, and invoked for analysis in abstract terms, because the very attempt to do that contradicts the relation in which we stand to that real.

God is not that which can be assigned by us to any class or set, or proposed for passive measurement. It is very similar to the mistake that people make when they label another person without any attempt to get to know them as a person. You cannot objectify a person—or if you do, then in so doing you immediately cut yourself off from any chance of engaging with them as they really are.

In order to begin to grasp the truth contained in the next paragraph, one must first meditate long and hard on this point about objectifying a person.

With God this goes further. You cannot say God is an 'entity' (still less a 'hypothesis')—or if you do, then in so doing you immediately cut yourself off from any chance of engaging with God as God really is.

There is something pathologically misconceived about phrases such as 'God is a hypothesis', 'the theist hypothesis', and 'the God hypothesis'. This is about usage. There is an attitude which simply assumes from the outset that God can be regarded as a means to an end, and this attitude is misconceived—not just inappropriate but utterly failing to grapple with the truth of the situation.

To be clear, I am not saying one cannot think about the widest context of our lives in an abstract way, but it won't help to adopt ways of speaking which approach that context as another thing alongside things. Such an approach would be nonsensical, like trying to contain a three-dimensional space in a one-dimensional line. I am also not saying that one cannot think about people in an abstract way, or discuss their characteristics. My point is that one cannot think that way about their very right to be acknowledged as personal, without actually taking the step of proffering to them your willingness to recognize such a right, and opening up the beginnings of a relationship on that basis. Without such willingness, the discussion is empty.

Suppose that someone sets for themselves the goal that they be better informed of the reasons why there is a physical universe. They might contemplate putting God in the role of providing a cause. However, if such a person desires to remain entirely in charge of their own identity and to feel little or no obligation to acknowledge God beyond the role of jigsaw piece, then they have misconceived what the word 'God' refers to. They have misconceived so much, I think, that really they are using the word 'God' to refer to that which is not God. So no wonder their programme fails to satisfy. It is a nonsense programme.

The effort of trying to think coherently about God is very much like trying to grapple with things such as goodness, or the good, and things like mercy, justice, and compassion—things that elude analysis, and that a piece of music or a simple human gesture might capture or express just as well as, if not better than, a scientific or a philosophical paper. Dawkins starts out by saying 'let's see what can be expressed in analytical language'; theology starts out by saying 'let's help one another think more clearly about what lies beyond such language, but that we sense nevertheless, right in the core of our being.' Wittgenstein[82] is helpful in getting this distinction clear, and it can be found also in St Augustine of Hippo and St Thomas Aquinas, among many others. It has been affirmed by C. S. Lewis, Herbert McCabe, Donald MacKinnon, and Rowan Williams in recent times (to name a few).

The first step in theology is to become aware of the impossibility of making God the subject of critical analysis. This sounds like an abandonment of any hope of ever saying anything useful about God, or of comparing notes, but it is not. It simply means that we have to look up, not down, when trying to speak of God. The rest of this chapter is an attempt to clarify what this means, but the reader must not expect the kind of clarity that can be captured in a syllogism, because that is not on offer. Instead, I will offer a sequence of examples which show how thoughtfully religious people in practice approach the task of speaking of God and responding to God. It is by observing how a word is used that one comes to understand what people mean by the word. And in order to get a growing sense of the answer to the question 'what is God?' one must at every step also take a serious interest in the answer to another question, namely, 'what modes of response to God are appropriate?'.

> A question has wedged itself between his learning
> and his awakening: how does one map a place
> that is not quite a place? How does one draw
> towards the heart?
>
> *Kei Miller*[44]

14.5 Four Witnesses

In the following we will address the issue of modes of response to God. We will introduce four examples or witnesses: Jesus of Nazareth as he is described in the gospels; Dietrich Bonhoeffer; the Desert Fathers (as representatives of a wider phenomenon); and the poet R. S. Thomas. In each case the mode of response indicates, in some measure at least, what it is that people mean when they say 'God'. But possibly the more important lesson is the mode of response itself. Instead of agonizing over the question 'what is food?' sometimes it is wiser simply to eat the food.

14.5.1 The Gospels

Like all arguments, the arguments offered by Hume and Dawkins involve various tacit assumptions. Some assumptions can safely be left unspoken; for example, the assumption that uncontroversial words have their usual meanings unless it is expressly stated otherwise. However, the tacit assumption here is central. Both writers assume that when we approach the fundamental reality of our lives, what we are looking for is explanations of how things work or how things came to be. To put it a bit over-simply, Hume implies that human beings are, at the heart of their being, philosophers, and Dawkins implies that they are scientists. This is to be contrasted with the assumption made by Jesus of Nazareth, that human beings are, at heart, people in need of forgiveness and a fresh ability to live better lives. What they are looking for is mercy and encouragement.

In Hume, and in Dawkins, titles such as 'Supreme Being' or names such as 'God' are introduced as concepts that can be passively looked upon and turned this way and that with our reasoning capacity. They are hypotheses that we can entertain, asking ourselves the question, 'is this useful?'.

In the accounts called gospels, when you hear Jesus of Nazareth at work, something else is going on. He never uses a word such as '$\vartheta\epsilon os$' ('God') as an abstract hypothesis that might or might not be useful for analysing the structure of physical events and logical arguments. And it is astonishing how often he uses another phrase, largely introduced by him, namely the one commonly translated 'your heavenly father' ('$\pi\alpha\tau\grave{\eta}\rho\ \dot{\upsilon}\mu\tilde{\omega}\nu\ \dot{o}\ o\dot{\upsilon}\rho\acute{a}\nu o\iota s$'—*Pater hymon ho ouranios*—'Father of you the heavenly'). That translation is rather unfortunate nowadays, because the word 'heavenly' is liable to be misunderstood. It should not be understood to mean 'located in a dubious never-never land'; it should be understood to mean 'absolute' or 'unbreakably so; defining'. It is a contrast with 'earthly father', a way of saying 'parent, but not your earthly parent; your other one, your ultimate parent'. The phrase 'your heavenly father' has the connotation: 'the absolute reality which both calls you and sources you'. This is the foundational reality whose nature is continually being more fully expressed as the universe develops. My point is that Jesus does not invite us to inspect a concept that we might or might not find useful. He says he is telling us about the source of, and creative pressure on, who we are, whether we have realized it or not. He invites us to seek, ask, mourn, forgive, come to our senses.

And he begins by announcing, 'think again' (the idea captured by the somewhat imperfect translation, 'repent'). He means 'change the way you think', not just 'change what you think'. The absolute originator and inspirer is a helper to be asked, not a servant to be summoned. Hume writes words such as 'Supreme Being' and 'Author of Nature', but he is not writing about a helper he has asked; he is writing about a servant he proposes to summon (and then dismiss). It is the very way of writing, and the way of thinking behind it, that is faulty. It is not completely faulty: it is a marshalling of reason and careful argument, and that aspect we should respect. But it invokes words that seem to be referring to the reality whom Jesus of Nazareth and others prayed with, experienced, and demonstrated, but which in fact refer to some construct of Hume's imagination.

This sounds like I am being unfair to Hume, and I think he would reply by insisting that I express more clearly what it is that I am referring to when I refer to God. This is very much what the *Dialogues* are about. My reply, and it is the reply that I think theology has to make, is to agree that theological talk quickly becomes meaningless if it is presented purely in the mode of analytical talk. But that is not the only mode in which we have to speak, as human beings. The combination of labels, analysis, and passively summoning up concepts to assess and dispose accordingly, does not give access to all of reality. It does not give access to friendship, for example. Nor is this the right way to express an adequate or appropriate response to all of reality. In particular, the sharing of truth about the nature of God is not done that way. According to our first witness—the gospel accounts—it is done another way, through parabolic stories, extraordinary encouragements, daunting challenges, and wonderful actions.

14.5.2 Bonhoeffer

The Lutheran pastor, theologian, and dissident Dietrich Bonhoeffer* is one of the more significant Christian voices of the twentieth century. In his *Letters and Papers from Prison*,[14] Bonhoeffer is helpfully forthright about past and present attempts by Christians to apply theology in misguided ways:

> The movement that began about the thirteenth century (I'm not going to get involved in any argument about the exact date) towards the autonomy of man (in which I should include the discovery of the laws by which the world lives and deals with itself in science, social and political matters, art, ethics, and religion) has in our time reached an undoubted completion. Man has learnt to deal with himself in all questions of importance without recourse to the 'working hypothesis' called 'God'. In questions of science, art, and ethics this has become an understood thing

---

* The spelling is correct; the family name became Bonhoeffer with no umlaut before the nineteenth century.

> at which one now hardly dares to tilt. But for the last hundred years or so it has also become increasingly true of religious questions; it is becoming evident that everything gets along without 'God'—and, in fact, just as well as before. As in the scientific field, so in human affairs generally, 'God' is being pushed more and more out of life, losing more and more ground.
>
> ...
>
> The attack by Christian apologetic on the adulthood of the world I consider to be in the first place pointless, in the second place ignoble, and in the third place unchristian. Pointless, because it seems to me like an attempt to put a grown-up man back into adolescence, i.e. to make him more dependent on things on which he is, in fact, no longer dependent, and thrusting him into problems that are, in fact, no longer problems to him. Ignoble, because it amounts to an attempt to exploit man's weakness for purposes that are alien to him and to which he has not freely assented. Unchristian, because it confused Christ with one particular stage in man's religiousness, i.e. with a human law.

In quoting Bonhoeffer I don't intend to imply that I completely accept his analysis of what he calls the 'adulthood' of the world, but he is right to tell Christians to stop being ignoble, and he goes on to provide helpful pointers in the difficult task of saying what sort of talk theological talk is. What sort of sense does it make? Here is an example of Bonhoeffer on this:

> The difference between the Christian hope of resurrection and the mythological hope is that the former sends a man back to his life on Earth in a wholly new way.... The Christian, unlike the devotees of the redemption myths, has no last line of escape available from earthly tasks and difficulties into the eternal, but, like Christ himself ('My God, why hast thou forsaken me?'), he must drink the earthly cup to the dregs, and only in his doing so is the crucified and risen Lord with him, and he crucified and risen with Christ.
>
> ...
>
> Redemption myths arise from human boundary-experiences, but Christ takes hold of a man at the centre of his life.

(By 'boundary-experiences', Bonhoeffer means issues that people do not in fact care about very much, such as the meaning of the wider universe or life after death.)

Also, in another example:

> Our relation to God is not a 'religious' relationship to the highest most powerful, and best Being imaginable—that is not authentic transcendence—but our relation to God is a new life in 'existence for

others', through participation in the being of Jesus. The transcendental is not infinite and unattainable tasks, but the neighbour who is within reach in any given situation.

These comments were addressed to Christians, so they are not framed in language designed to help an onlooker find a way in. I include them here in order to give an impression of what Christianity can bring to the table. Bonhoeffer acknowledges that the way of thinking that is involved when one invokes a 'working hypothesis' called 'God' proves to be empty in all areas of life. This is so in areas such as ethics and origins, as well as other areas (I think this is what he means by 'it has also become increasingly true of religious questions'). But this does not imply that God is empty of meaning for Bonhoeffer. He considers that the avenue he rejects (in implicit agreement with Dawkins) not only does not exhaust all possibilities, but also was never central, nor what Jesus stood for.

Note that there are two senses of the word 'Christian', and both appear in the quotations above. The first sense is 'those people who self-identify under that label', and this is a group which Bonhoeffer identifies with but also critiques. The second sense is 'a person who really gets to grips with what Jesus was showing us, and tries to live it out', and this is what he wants to advocate. What he describes as 'our relation to God' is not an attempt to solve philosophical puzzles or answer impenetrable questions, but is rather the commitment to embark on 'a new life in "existence for others"' which he sees as 'participation in the being of Jesus.' Obviously this is shorthand for a theological idea which would take longer to describe more fully. In his letters and other writing Bonhoeffer uses the word 'God' as it is used by people at the interface of scholarship and pastoral care. He can draw on a large amount of learning, including history and psychology and so on, but in the end he wants to know how it pans out in terms of how people live; this is where he looks to find the meaning of religious language. And this is where he looks to contribute in his turn. The word 'God', in his experience, is the word which signals *that fullest reality which a human life, or any other life, never completely expresses, but can reflect in some measure.* God is the reality we are reaching for and joining with when we live life wholly and well, responsive to the needs around us and enacting the good that can be enacted by our efforts. That, at least, is what the present witness (Bonhoeffer) affirms.

Dietrich Bonhoeffer was quite aware that the desire of atheism, at its best, is to affirm this same type of life but without religious language. In view of this, why speak of God? How does it help? We speak of God, rather than a collection of good items such as values and aspirations, in order to recognize, correctly, that our goal exceeds what we have so far, as a community, managed to even think about, yet alone enact, and also in order to recognize that it is not just about our trying to do better; it is as much or more about God opening us and helping us to do what would be impossible without God's help. Finally, we speak of God in order to recognize that in this journey we are, as a matter of fact, called out of our own orbit into a loving relationship. Hard to pin down as such words are, that is the fact of it, in the experience which Christian witnesses affirm.

Bonhoeffer speaks of 'drinking the earthly cup to the dregs'. This is a powerful, gritty, intellectually thorough commitment to engagement with the world. Abraham Lincoln had some sense of it, I judge, and so did Rosa Parks.[49] The great physicist James Clerk Maxwell showed how it can play out in a life in science.* 'Participation in the being of Jesus' expresses Bonhoeffer's sense of how it 'works', how we manage to live more fully for others than we could otherwise manage to do. We become part of an expression of creative goodness that takes place in the ordinary everyday world of mud and stone, grass and air, families and colleagues. In this way our humanity is expressed appropriately and fully. This is a partnership in a difficult task, but one in which we are offered support in our innermost being.

I asserted already that analytical discourse, which is what science is all about, is not the only mode in which humans have to speak. Now I will add: nor is it the most important mode. The most important mode is the way we live, and how we give ourselves, which is indicated especially by what we give up or go without as we falteringly attempt a full expression of our human personhood. This is what Bonhoeffer was talking about, and this is what Jesus of Nazareth was showing us how to do. I will return to this after presenting my two further witnesses.

---

* 'Happy is the man who can recognize in the work of Today a connected portion of the work of life, and an embodiment of the work of Eternity.' J. C. Maxwell, note written in his twenties. 'I think Christians whose minds are scientific are bound to study science that their view of the glory of God may be as extensive as their being is capable of.' J. C. Maxwell, unfinished draft discovered after his death, quoted in L. Campbell,[17] p. 404.

14.5.3 The Desert Fathers

Towards showing what many Christians at least mean by God, all parts of the Christian tradition contribute and balance one another. When studying the analytical challenges, balancing factors come from the contemplative and mystical traditions. These have been helpfully reasserted in recent years by the Trappist monk Thomas Merton and the Benedictine John Main, among others, and by a rediscovery of the Desert Fathers of fourth-century Egypt. Rowan Williams begins his introduction to their legacy with a statement that has a striking resonance with the quotations from Bonhoeffer above:[79]

> One thing that comes out very clearly from any reading of the great monastic writers of the fourth and fifth centuries is the impossibility of thinking about contemplation or meditation or 'spiritual life' in abstraction from the actual business of living in the Body of Christ, living in concrete community.

The Desert Fathers were responding to a deeply felt longing for truthful human life, which learns from history and from the demonstration offered by Jesus of what this involves. They did not try to argue points about the origin of the universe. Rather, they tried to live in the context of an absolute challenge to perfection of the soul, combined with a complete and serious insistence on the reality of forgiveness. This, they argue, creates the possibility of honest and constructive communal life. It is what liberates and emancipates humanity. Another basic lesson emphasized by the mystical tradition is that we mistake the nature of God when we take it that we can 'think' God in the way that we can entertain passive propositions. '[God] may well be loved, but not thought. By love may he be gotten and holden; but by thought never.' (*The Cloud of Unknowing*, Ch. 6).[6] Reason is an important ally, and on the basis of cautious, humbly sceptical enquiry we can journey right to the brink of engagement with God. But that engagement is not and never can be passive, like entertaining a proposition. It is an engagement of our person with another who cannot be thought, only welcomed; cannot be known of, only known; cannot be owned, only shared. It is not comfortable (it is especially not that), but it is fulfilling.

14.5.4 Soundings from R. S. Thomas

We insist, and it is important to insist, that this kind of paradoxical or deliberately oblique language is not an abandonment of sense.

Something constructive passes from one person to another when we speak this way, and something goes on in the self when one thinks this way. An exploration that has helped many in recent times, and that I think holds lasting value, is the one carried out over an extended period in the poetry of R. S. Thomas.[57] Thomas reports from his voyages over a territory hard to map, and which perhaps cannot be mapped, but which many of us recognize. His work is important because it is uncompromising in its honesty, serious about science and the modern world, and single-minded in pursuit of truth, but determined to address what is learned in silence as well as what is learned in speech. Also, one feels in his case that one is engaging with a sharp intellect as well as a craftsman and an artist. He is the intellectual equal of philosophers such as Hume and Nietzsche, for example, but one who chose a different path, fleeing academia in search of something academia does not handle very well. Atheism is squarely contemplated in his poetry. He presents it sometimes starkly, in a first-person voice, and this will puzzle some readers, but this is not a denouncing of all religion, only a denouncing of much of what passes for religion. It is a grappling with the difficulties of faith. For Thomas, atheism stands as an option, and indeed as an escape, because it offers an escape from the painful struggles of theism, but he continually finds it unpersuasive, and he prefers the struggle.

Thomas is mostly unsettled, not settled, in his searching thought, so it is important to hear the later as well as the earlier poems when learning from him. He is helpfully frank about the unpersuasive nature of passive religious argument: 'The mind's tools had / no power convincingly to put him / together' ('Perhaps', in *Frequencies*[68]). This is an agreement with Hume and Dawkins, up to a point. However, he repeatedly insists on his intense and demanding experience that there are forms of truth that are more than, and more objective than, mere human feeling, but which elude critical analysis. 'It is its own / light, a statement beyond language / of conceptual truth' ('Night Sky', Frequencies); 'this great absence / that is like a presence' ('The Absence', Frequencies); 'one not to be penned / In a concept, and differing in kind / From the human; whose attributes are the negations / Of thought' ('After the Lecture'). 'But the silence in the mind / is when we live best, within / listening distance of the silence / we call God.' (in *Counterpoint*[71]). Silence is absence of sound, and it here signifies absence of most forms of communication, but it does not signify the complete absence of anything whatsoever that can be attended to.

A poem called 'The New Mariner' (in *Between Here and Now*[69, 72]) opens, tellingly,

> In the silence
> that is his chosen medium
> of communication and telling
> others about it
> in words. Is there no way
> not to be the sport
> of reason?

The way of looking, and of communication, that the poet is grappling with is not well-matched to words, and it is not the sport of reason. It is something else; but that does not make it unreasonable. It is an objective part of human experience that refuses to lie down under accusations of delusion or confusion. Is it a way of looking worth sharing? Decidedly, yes. It sometimes flowers into explicit meaning, for Thomas, and even when it refuses to do that, it has unmistakable effects. It carries out its soul work. 'Night Sky' concludes, '… Every night is a rinsing myself of the darkness / that is in my veins. I let the stars inject me / with fire, silent as it is far, / but certain in its cauterising / of my despair.' This is not nature mysticism, but the experience of something hinted at by nature, or incompletely expressed in nature, and the use of nature to provide metaphors. In all Thomas's work, the presence, or the whatever-it-is that is like a presence, is not mere resources at our disposal, but is charged with a more personal challenge, 'like intentions among suggestions.'[48]

The poem beginning 'But the silence in the mind' continues 'It is a presence, then, / whose margins are our margins; / that calls us out over our / own fathoms. What to do / but draw a little nearer to / such ubiquity by remaining still?'. On the face of it, nothing has happened in the 'drawing nearer' proposed here, and yet something has happened: a wonderful, peaceful, consoling, and creative something. Atheism will interpret all this in psychological terms, of course, but as soon as you buy that interpretation you have destroyed the very drawing-near that was being meekly appropriated.

One of the important conclusions at which Thomas arrives, and I think that this most unsettled poet found it possible to settle here, is again reminiscent of the point I have quoted above from Bonhoeffer and from Williams on the Desert Fathers. It is to come back to 'everyday life', to 'plain facts and natural happenings' and there discover the absolute reality and compassionate engagement spoken of by Jesus. The poem 'Emerging' (from *Frequencies*[68]) is one of his most important in this respect. 'The mind, sceptical as always / of the anthropomorphisms / of the fancy, knew he must be put together / like a poem or a composition / in music, that what he conforms to / is art.' He concludes,

> We are beginning to see
> now it is matter is the scaffolding
> of spirit; that the poem emerges
> from morphemes and phonemes; that
> as form in sculpture is the prisoner
> of the hard rock, so in everyday life
> it is the plain facts and natural happenings
> that conceal God and reveal him to us
> little by little under the mind's tooling.

'It is matter [that] is the scaffolding of spirit'. That is, perhaps, the central claim of Christian thought. God is revealed in the very place that conceals Him. Conceals, because that place consists of just ordinary facts and natural happenings. Reveals, because acquiring a full receptivity to those ordinary natural things is how we see God, and this is the only way He can be seen.

The reader will suspect me of hereby making God redundant. I already said that if you think God is useful, think again. God is no more useful than is beauty useful, or justice, or compassion, or brotherhood. But this is not to say God does not actively participate in the ongoing development of the world; God participates, I think, by offering an influence appropriate to what is involved in all personal interactions, which is to say, subject to a veto on our side, and respecting our autonomy. I will say a little more about this in Chapter 21.

14.6 The Refutation of the Superfluity Argument

The main business of the chapter has now finished. I have written enough to show that authentic Christian witness—that is, a response which tries to grasp and join in with the way of life introduced by Jesus of Nazareth—does not involve superficial religiosity and does involve correctly constructed thinking. The chapter has made some excursions into analogy and illustration in an effort to be clear, so it may be useful to state now the central academic point that has been achieved. This is that, considered in purely intellectual terms, the argument I have called LEP fails owing to a bad premise. By putting God into the role of 'explanatory hypothesis' it has already denied the possibility that the appropriate response to God is akin to friendship. Thus the argument has assumed into its premises the thing that is claimed to be the conclusion.

More generally, I think what we can learn here is that it is not possible to come up with some well-constructed argument which tells you that your true and fullest context has no interest in who you are. Nor is it possible to deduce that such a conclusion is highly likely to be right, or even quite likely. The whole way of arguing is faulty and proves nothing at all one way or the other. In this area it is not possible simply to reach for the 'off' switch while retaining one's intellectual integrity.

Airy and safe academic talk about God quickly degenerates into a sort of word game which doesn't help. That is why I have drawn on human experience and human effort. The above witnesses are witnessing, I maintain, to something that results in creative human life, and their witness has recognizable common themes, and it is a witness to something not captured in the argument I have called LEP. These witnesses use the name 'God' for that which challenges and restores the heart of their identity, rather than for an explanatory concept which might or might not be useful. So LEP fails, not on the basis of being unsound in its own terms, but on the basis of simply failing to grapple with God.

I don't want to dismiss or be at all disdainful of the genuine goods which are championed by atheism. These include the intense and noble desire (I almost said holy desire) to avoid forms of religiosity which lapse into lazy or superficial or superstitious thinking, or which get hung up on unanswerable questions, or which make people manipulate others by preying on their trust or their guilt or their fear. Equally, though, I want to insist that the kind of responses and attitudes that I have discussed in this chapter are humane and reasonable and they do not deserve disdain. Nor are they artificial add-ons to human life, optional extras that we can well do without. Rather, they are ways of allowing us to be human in full, finding a more complete encounter with plain facts and natural happenings, such that our instinct for hope and meaning is well directed.

At the end of Section 14.4.1 I asked the rhetorical question, 'what could possibly be wrong?'

What can possibly be wrong, and is wrong, is the attempt to approach all learning by only employing categories and ways of thinking that leave you fully in control of your own learning process. What is wrong is to think that every kind of truth can be attained by making a definition of some X and perusing the X dispassionately.

The superfluity argument, or LEP, takes as premise the following assertion:

> Premise: the only kind of figure or reality that can have intelligent understanding and intention is a kind that is to be thought of as an entity and can be spoken of in the role of explanatory hypothesis.

LEP succeeds in showing that it is untenable that such an entity can play the role of final cause or adequate explanation for anything. But the premise should be rejected, because of the following considerations:

1. We do not know it to be true.
2. The very idea that one can consider as a 'hypothesis' or a 'useful idea' the notion that another is capable of, and interested in, personal relationship, is itself misconceived. As long as that possibility is being treated as a hypothesis, it is in fact already being denied. Therefore this approach cannot serve as the beginning of an argument to deduce that the possibility should be denied.
3. The experience of living as people who form personal relationships teaches us this, and also shows us that, nevertheless, personal relationships can grow, because they grow by steps of trust and attention, and investing of one's own identity in another.
4. The evidence of the well-considered experience of humankind is that we can encounter our larger context both in impersonal and in personal terms.

The phrase 'well-considered experience' here refers to the sum of human experience that has been thoughtfully shared and discussed over long periods in communities in which there are signs of human flourishing. This includes the contribution offered by mystical experience, but more importantly it includes the whole rich tradition of thought and action that attempts to enact in the world an appropriate response to the notion that we are called to be 'children of God', understanding that God is whatever it is better to be than not to be.[7]

As Bonhoeffer notes, this does not invite us to fantasize about the 'best Being imaginable', but rather to pick up the notion of 'existence for others' and try to inhabit it, make it a happening that actually happens, while recognizing that we are utterly incapable of doing that unaided, but the parables and other encouragements of Jesus tell us broadly how it works.

Items 1 to 3 above suffice to point out the fallacy in LEP and in similar arguments. In the next, brief, chapter I will elaborate on item 4.

15
Drawing Threads Together

> This continues the theme opened up by the previous chapter, by showing how religious experience operates and how it informs human life. This includes the experience of mercy invoked by music, for example, and the notion that we are fundamentally called into a loving relationship, and given a role to play. This is not about providing further components to scientific explanations, but it does act to encourage us to practise science.

In the previous chapter my main aim was to bring to light what is wrong with a certain attempt at reasoning, or type of reasoning. It was not a case of a false step within a sequence of connections, but rather the fallacy of going about things the wrong way, misconceiving what sort of area one was in. In order to show this I mostly tried to show what the nature of the territory is, first by considering what is true of relationships that involve steps of opening and sharing of personal identity, then by bringing in various witnesses, and, throughout, by looking at the innate nature of mutual trust.

In the present chapter I will comment further on the evidence that emerges not just from the witnesses I have named, but more generally from widespread human experience. It will not be my aim to make an irrefutable argument for the way the evidence points (I think that cannot be done), but rather to show what sort of evidence it is, in the sense of what its categories are, and hence what sort of reactions appropriately address it.

Many people consider that that which supports the universe or enables it to be and to become is thoroughly impersonal and hence not a suitable focus of attitudes such as trust or friendship. Fair enough. But they should not claim that people who do think that physical reality springs from, and is called into fullness by, a dependable, non-contingent, fair, and compassionate Caller assert this because they think it gives a simple account of the origin of organized complexity.

Admittedly, that is how some people advocate it. But that is not what is going on in most theology, philosophy of religion, and the witness of those faithful who have the kind of intellectual gifts that enable one to analyse one's own thought cogently. No, what these people mostly think is that, simple or not, this does indicate truly the ultimate origin of organized complexity, but more importantly it does a lot more besides.

The phrase 'simple or not' needs comment, since it touches on the heart of the issue (and it is liable to act like a red rag to a bull!).

When I hear Mozart's Requiem, the quality that I am chiefly engaged with, and appreciative of, is not its simplicity but its beauty. Similarly, when we approach God, a whole range of considerations come together, including the reasoning about origins, but not chiefly that, and certainly not stopping there. The range includes historical evidence, and the experience of beauty and of forgiveness and of loss, and the sense of having been called out of oneself to become someone better, and that of being a part of a body whose imperfect members express love, and the sense that the only proper location where our standard of values can be placed is at infinity. It is a throwing of one's loyalties in a direction sensed to be profoundly valuable though not fully understood. This involves gaining hope and encouragement in a situation of imperfection and compromise, in such a way that one is thrown back into that imperfection and compromise with the task of joining a shared effort to improve it. Overall it is a way of understanding human identity that has that combination of beauty and goodness that we have learned to associate with truth.

If this were primarily about scientific explanation, then simplicity would be a primary concern. But since it is primarily about a range of other things, its beauty does not consist primarily in the kind of simplicity one looks for in scientific arguments. This is not a solution to any scientific puzzle. Insofar as it has any impact on science, it functions as an invitation to pursue science. A good parent encourages us to think and is not out to trick us.

This way of understanding human identity is not about hazarding guesses about the origin of the universe, nor is it about expending energy in a fruitless attempt to answer unfathomable questions. It is about acting today out of a willingness to entertain the thought that we are fundamentally persons whom Love meets. This provides hope on one hand, and passion on the other—passion to realize a just world, that is (and also the danger that this passion will get misdirected). It provides these things, and can only provide these things to a reflective person, if it also satisfies one's conscience, including the duty that is owed to rationality. It meets that duty in part by involving an appropriate epistemic humility.

We experience and respond to the mercy that liberates, communicated to us by a piece of music for example, but we do not pretend to know the essence of the source of that mercy. Some of us think Jesus was raised from death, because that is what was and is experienced, but we know he might not have been, and we also know that we are not quite sure what we are talking about when we say that he was. We are aware of the acute dangers of coercive religion, but we note that the idea that all are valued, and equally valued, by a perfectly fair and loving Parent, has been the single most liberating insight that has ever taken root in the human psyche. Or at least, a good case can be made for this, because it has both given people meaning in the immediate situation of their joys and pains, and it has also underpinned the movement away from ancient religious power structures to the modern secular democratic principle (see for example Sidentop 2014, Williams 2005, Sacks 2012[63, 80, 58]).

Saying what religious response is about is highly non-trivial, and it cannot be done by means of definitions. If we start to line up terms and titles such as 'Author of Nature', 'Transcendent Creator', 'Love, liberty, equality, fraternity', 'Mathematics', 'Trinity', 'Heavenly Father' and regard them as definitions of a concept, and then peruse the concept, then it is not the titles but our attitude that is the centrally defining aspect of what it is we end up considering. We end up considering that which can be considered this way. And that which can be considered this way is not the totality of that which is real and can command our allegiance. So the question is not, 'which of these titles is best?'. The question is, 'how do we gain access to the rest of reality, to that which is missed out when one adopts the analytical stance?'. This can be approached, I have argued, in the kind of way I have sketched, by paying attention in certain directions, and allowing what is attended to to have an influence on oneself. Therefore in this and the previous chapter I have given examples of how it works out in practice, alongside paying due care and attention to the possible structure of the thought involved.

It is the use of personal language that is at the heart of the issue regarding atheism and theism. Non-theism (i.e. atheism and agnosticism) can usually accept abstract moral principles as guidelines for human life, and non-theism can accept a somewhat mysterious origin or support for the universe as long as we leave it there: just say the properties are unknown to us. Or, it might be allowed, that which is real without being composed of physical stuff could have the kind of attributes that impersonal abstractions have: logic and mathematics, perhaps, but not personal qualities such as the ability to have intentions and opinions. Meanwhile, the 'absolute parent' described by Jesus is to be understood in personal terms: one who can know and be known, trust and be trusted, be thanked and thank, receive and express anger, suffer, hope, grieve, rejoice.

How credible is this? asks the modern-day atheist (very much aware of the fruitfulness of the scientific method). But the prior, and in practice more important, question is, 'how should I go about assessing its credibility?'. Hume invites me to ask myself, 'does it settle various philosophical dilemmas about the origin of mind?'. Dawkins invites me to ask myself, 'does it settle various scientific difficulties about the origin of life?'. The Christian community invites me to ask myself, simply, 'are you willing to be loved?'. And 'are you willing to join a community that tries to learn and to show what this might mean, involving as it does the attitudes and experiences that Jesus spoke of when he described "the Kingdom of God"?'.*

An important and helpful thought in all this is that that which most fully deserves our allegiance is not to be approached as *useful*. I have mentioned this already; I repeat it here because I myself have found it to be a very helpful thought. The parent spoken of by Jesus cannot be *used*. That is not how the relationship works. God is not a scientific hypothesis that is useful for explaining things, or for finding simple ways of looking at things. Rather, God provides the very concept of explanation, an intelligible universe, and the urge to explain. The answer to the question, 'what has theology ever offered to science?' is: 'science.'

So God is not an element in a description which seeks to explain things in terms of other things; rather, God expresses the truth of what

---

* The phrase 'Kingdom of God', being translated into modern English, means, approximately, *the domain of generous and mutually supportive relations*; a place where people discover themselves to be valued and in which their more generous instincts and most penetrating insights can flourish, like seeds finding good soil. The term has, however, been widely 'religionized' to mean something else, such as membership of a church.

things are in themselves: their meaning and role. He also provides the kind of sympathetic wisdom that is needed in situations of tragedy, when the urge to explain has to be set to one side because random disaster has no meaning in what led up to it. Such situations are ones to which meaning has to be brought by the way we respond.

Dietrich Bonhoeffer was imprisoned for his opposition to Nazi policies in Germany. A guard at one prison offered to help him escape and 'disappear', but he declined, in order to avoid Nazi retribution on his family, some of whom had already been incarcerated. Bonhoeffer's final place of imprisonment was Buchenwald concentration camp, where he was a courageous and reassuring presence for other prisoners, and performed at their request a short Sunday service the evening before he was executed on Hitler's orders. The attributes of God are demonstrated not by a philosophical argument whose conclusion they are, but by the way their acknowledgement by us both motivates our best endeavours and brings light to our darkest places.

l'engrenage

I climbed up wearily, tread upon tread,
word on word,
spiralling stair of argument,
because I was told to look
in the boardroom on the umpteenth floor.
The windows grew large.
There were distant views, crisp and airy,
but the rooms got more and more empty.
Minimalist, sharp designs, crystal-clean,
chairs not meant for rest.

The air grew thin.

You had to give it up.

But then I felt the thrumming of the building,
and saw it was a ship.
I began to explore below decks,
finding more and poorer people
the lower I went.
The hum in the railings
grew in the floor.
There were patches of oil,
and more and more ragged faces.
Then, opening a door on steel
and heat I found him
wrenching the gearing,
turning the fire's red to blood
in the engine-room where God breathes.

16

Extraterrestrial Life

> We overhear a conversation between a terrestrial speaker and a representative alien from a planet orbiting some other star. The conversation touches on moral and spiritual issues. The main aim is for the attempt not to be parochial.

Arguments such as the one of this book, and expressions of human experience such as that of the previous chapter, have to make their way in the larger context of the cosmos as a whole. That is, any way of framing human life in the round, or of saying what we should aspire to, will not be either honourable or credible if it can do it only by implying that planet Earth is some sort of Chief Planet of the entire cosmos, and ignoring evidence to the contrary. The point is that, whether or not intelligent life exists elsewhere, we have to take seriously the fact that it might.

There is, in my opinion, a good chance that life exists somewhere in the universe in addition to on planet Earth. This opinion is admittedly only a hunch based on some preliminary data. The vast numbers of stars in the universe already made many people guess that there might be life elsewhere, and with the data from the now famous *Kepler* space telescope, we have very good reason to believe that there are large numbers of planets in our own galaxy that are located in the so-called habitable zones of stars, where there is a possibility of things like liquid water and not too much ultraviolet radiation. It seems reasonable that there may be life out there.

I would also hazard a wild guess that there is developed, complex life, even perhaps conscious life, somewhere else. This is much more of a guess; really I don't know and nobody knows. But it is interesting to think about what it might imply if there were.

This chapter offers a few thoughts that don't depend on whether or not there is conscious life elsewhere. It is sufficient that we may suspect there is. Because if we may suspect there is, then we have to think of our own position in the cosmos accordingly. That is, we are not central but we are part of it. Everyone already knows this in relation to the physical cosmos; we have to know it in relation to aesthetic, moral, and spiritual realities too. The reader may think that religion is terribly out of date here and didn't give much thought to this. In fact it has been discussed on and off for centuries, but there isn't really very much that needs to be said.[78]

It seems to me that if we encountered complex life on another planet, then it would not be too hard to detect the presence of conscious reflection if it were also there, as long as we were carefully looking for it. And once we had detected that, and shown signs of it in ourselves, it would not take too long to develop a common language. The situation is not so very different from the one faced by explorers to unknown territory on Earth in previous centuries. Then there was a common environment on Earth to fall back on; here there will be a common galaxy and basic laws of nature to fall back on. And once we had a common language, some fascinating conversations would undoubtedly take place.

The initial situation would, I suppose, consist of passive observations—telescopes, etc. Either they would be gazing at and listening to* us, or we would be gazing at and listening to them, or both. But humans are a talkative lot, so I suppose we would want to send messages almost as soon as we could. And maybe they would feel similarly. Communication across the light years of space is a slow business, so it would not be like an ordinary conversation. We would just start sending information, and so would they, and after decades or centuries the message would reach the recipient. The replies would be added to the ongoing data stream and thus only after a long time would we or our descendants receive back the reactions. Still, those reactions would be of such great interest that I think we would certainly do it if we could.

Usually people of a scientific bent assume that the initial messages will consist of simple sequences such as prime numbers in binary code, in order to give evidence that can be easily interpreted. It isn't all that important what the initial messages are, as long as they can be recognized as having some sort of structure that gains attention. I think it wouldn't be long before we were sending a few tunes such as 'twinkle,

---

* I mean 'listening' in the sense of interpreting and reflecting on what is seen.

twinkle, little star' down the line, and soon after that something more ambitious such as a Mozart concerto. We would insert some questions about cell biology and quantum gravity as soon as we felt confident the questions could be understood, but before that we might well be transmitting video images of life on Earth, and images of ordinary people at work and play, having a meal, talking, smiling, frowning. In fact there would be a very great amount of argument about what we should and should not transmit; it will not be decided by a few scientists but by a much larger and more representative group of people.

I think that one thing I would want to say early on, and which in a way one was saying already just by starting up the transmission, is 'we see you not as aliens but as siblings, children of the same greater universe of things and values, and we want to trust you.' Another thing that would eventually appear in the conversation, after agreeing the elementary stuff such as mathematics, is simple propositions about behaviour. We would be reflecting on the miserable state of much of our own behaviour, while trying to put a best foot forward and behave as well as we could towards our cosmic neighbours. So we would want to be honest with them. I think that sooner or later we would send information about what our existence is really like, warts and all. We might be a bit coy about admitting the level of avoidable unfairness in human society, but it would emerge. And a lot of people would be desperate to say something about, and to ask something about, God; and in the interests of honesty and fairness, they would be invited to do so. And to avoid making fools of themselves, I hope they would think long and hard before working out what to say.

In the following I give an extract of how the transmissions might go, but, for the sake of simplicity, I have adopted the dramatic licence of making it like an ordinary conversation where you wait for the reply before saying the next thing. You have to imagine that an initial period of signing and language development has already taken place, so that we have developed a reasonable fluency in our mutual understanding, and can address sophisticated concepts. We pick up the conversation at a point where it turns to moral issues . . .

'We Humans have a saying, a sort of rough rule of behaviour, which says that you shouldn't do to another what you would not want them to do to you.'

'Oh yes, we say the same. Or better still, try to do for the other the sort of thing that you would like if they did it to you.'

'Oh good! We agree. But it's not always obvious what it means in practice.'

'No indeed. But we Keplians teach our young Keplians to use their imaginations, to imagine what it might be like to be the other Keplian, to be in their place. And then when the other benefits, they will feel that they have benefited too.'

'Yes! Only, to be truthful, we find this very difficult and often we don't act like that.'

'We have certain practices that can help.'

'Oh? Would you tell us about them?'

'Well, often before making decisions we sit quietly for a while and try not to think about anything in particular, and then we try to fall back, as it were, into our ultimate support, the one that never goes away, and we find there a sort of challenge and encouragement. It is a challenge that demands we be truthful, but also an encouragement that tells us we are precious. After putting into practice habits like this our decision-making gets better.'

'How interesting! And what else?'

'We find that meeting to share food and conversation together can help to build up positive links in wider groups. And we use food as a symbol too. It reminds us that what nurtures our inner life also comes from beyond us. We Keplians see ourselves as recipients through and through, in every respect; we are sharers, channels.'

'We are in each other's keeping.'

'Yes, and we also feed on %&% in our hearts by faith, with thanksgiving.'

'What?'

'Well, we think that our inner resources, our hope and our capacity for joy and for perseverance, and so on, do not reside permanently in us but have to be continually renewed, and it is good to remind ourselves of this, and be thankful that hope does get renewed, and perseverance does grow, somehow, renewed by a process we do not understand but benefit from.'

'Yes but it is an impersonal process isn't it.'

'Sorry, we're not sure what you mean.'

'What you called "%&%" just now is a shorthand for a complex array of physical processes.'

'No, it's a shorthand for what that array of processes amounts to in fact—what it is a realization of, an expression of.'

'But does it make sense to *thank* %&%? Can you also be angry at %&%? Can you ask for things?'

'Honest anger is OK. And bitter thoughts are better offloaded here than elsewhere.'

'And asking for things?'

'Don't you want to hear about all our gadgets and spaceships? Our neutrino furnace? Our Z-ray that can frazzle a snipjack at a thousand metres?'

'No, we want to know if you think it is legitimate to ask %&% for things.'

'In that situation you can only sqrxkz ask—that is the only asking; and you must try to name your sqrxkz need—that is the only needing; and then you will either be given that thing, or you will be given the task of going without it. But the art is to ask the very thing that a representative of %&% would ask.'

'Sorry, what is "sqrxkz"?'

'Sqrxkz is untranslatable. It is something like "absolute" or "complete" or "unconditional".'

'From what you just said, it seems that there could be no difference between asking and not asking: in either case, what you wanted to happen may happen, or may not.'

'Having the attitude of asking is itself a difference. But we think there is more to it than that. In the past the habits and assumptions of our forebears were often bad in ways they failed to see, but some among them asked for change, not just on behalf of themselves, but for everyone, and gradually it happened. Their gifts of hope and perseverance helped us all.'

'But we find that sometimes hope fails, and perseverance runs out, and there is no joy.'

'Yes, and then we are held by %&% in shared darkness, without the benefit of the gifts.'

'That is cold comfort.'

'We have arrived where abstract discussion does not comfort. But, nevertheless, what you achieve after hope has gone is tremendously precious; we give it our highest recognition.'

'May I raise a different issue? What do you do next when you have done something big that you ought not to have done? When your deliberate action has damaged another person, because you were lazy or bitter or thoughtless? How do you move on?'

'Here is a large issue. We feel that it involves justice and mercy in tandem, but forgiveness is ultimately the key. It will take a while to discuss, and maybe you can help us with this too. . . .'

There are several ways it could go. Maybe I have totally underestimated the sheer marvellous *otherness* of an alien species and the way they might think. But maybe not. . . . *We are more alike, my friends, than we are unalike.*[5]

Red Shift

Held by an image of our outer space:
Spots, dots, and whirls of white and red,
Time-tunnelling in silent grace,
Parsecs where only thought can tread.

Blue blazes of the younger fire,
Red smudges of the ancient mist,
Vast mergers of the flowing gyre
Down ages of the world persist.

These distant forms of space and truth
Work back upon the thoughts we frame;
Prayer puzzles through a shaping sieve:
Dead words or else a larger name.

Still, quietly ask the teeming sky:
Draws over there that which can love?
Lights there a dance which can convey?
Rests there a hold of things above?

17

Does the Universe Suggest Design, Purpose, Goodness, or Concern?

> The question is tackled of whether or not the natural world presents us with a picture empty of purpose, good, evil, or concern. No empirical evidence can entirely refute the claim that random fluctuation is the complete truth about the origin of all things, but it follows that this is not a scientific claim. Therefore it is a question of forming a reasonable judgement. It is suggested that the natural world has a depth and richness that exceeds what would be necessary for thinking brains to come to be realized in it. Also, notwithstanding the pain of the world, it is a project that merits our engagement and commitment.

This chapter addresses the question posed in its title. What kind of a universe have we got? Does it give intimations of a goodness, design, purpose, or concern at its source or in its foundation? Or does it suggest the opposite? Or is it empty of such intimations, either one way or the other?

To bring this into focus, and address a question which has been asked, I shall consider the following widely quoted paragraph:

> The total amount of suffering per year in the natural world is beyond all decent contemplation. During the minute that it takes me to compose this sentence, thousands of animals are being eaten alive, many others are running for their lives, whimpering with fear, others are slowly being devoured from within by rasping parasites, thousands of all kinds are dying of starvation, thirst, and disease. It must be so. If there ever is a time of plenty, this very fact will automatically lead to an increase in the population until the natural state of starvation and misery is restored. In a universe of electrons and selfish genes, blind physical forces and genetic replication, some people are going to get hurt, other people are going to get lucky, and you won't find any rhyme or reason in it, nor

Science and Humanity. Andrew Steane, Oxford University Press (2018).
© Andrew Steane. DOI: 10.1093/oso/9780198824589.001.0001

any justice. The universe that we observe has precisely the properties we should expect if there is, at bottom, no design, no purpose, no evil, no good, nothing but pitiless indifference.

— Richard Dawkins, *River Out of Eden: A Darwinian View of Life* (1995)[22]

I shall discuss the reaction to the natural world that is expressed in this quotation. I am mostly concerned with the last sentence, but let me first briefly comment on the opening that builds up the dramatic power. When you read the comment on suffering, it seems at first like a valid observation, one that 'sees through' the 'illusion' of the goodness of the world to all the harshness of 'the truth of things'. But think a little. If you had to write a couple of sentences in which you tried to capture a fair portrait of what happens in the natural world during the minute it takes to compose a sentence, would this be the portrait? Of course not. The suffering is not to be set aside, but it is less than half the story of most life, and it is less than half the story of life on Earth. Are all the careful, sympathetic, and fulfilling studies presented by naturalists such as David Attenborough, Gavin Maxwell, or T. H. White just some sort of rose-tinted spectacles and wishful thinking? No. Go and look in your garden, or in the forest, or the jungle, or in the river, or the ocean, or on the African plain. Is it the case that starvation and misery is the 'natural state' of affairs? Or are they part of a natural state of affairs which has here been grossly misrepresented?

These rhetorical questions are an appropriate response to a rhetorical piece of writing, but it is worth taking a moment to unpick this a little further. It is of course true that there is a huge amount of pain and suffering in the world, and all life comes to death. But there is also a huge amount of life itself, and death can only come where there is life. So it is not so obvious how this should be weighed up in a truth-seeking and balanced way.

Suppose there were just a handful of living things in some small ecosystem on a planet somewhere. Then one would not be able to claim that there are thousands of animals whimpering with fear during any given minute. There might be, for example, only one. Now take that same ecosystem and reproduce it several thousand-fold on different areas of that same planet. Now you can, if you want, assert that there are thousands of animals whimpering with fear during any given minute, and your statement would not be untrue, but it would be misleading. The point is not that we shouldn't raise an objection to the presence of fear and pain. Such things should be objected to. My point is merely that as soon as we admit that pain and fear is part of the mix, then there is going to be a lot of it if there is a lot of life.

The quotation I have mentioned is only a short extract, so it does not, on its own, present the full force of the argument behind it. However, I think the concluding sentence is a fair statement of where Dawkins and others are coming from, and it is the sort of thing you see emphasized by modern-day public speakers in the cause of atheism: 'The universe that we observe has precisely the properties we should expect if there is, at bottom, no design, no purpose, no evil, no good, nothing but pitiless indifference.' In the following I am going to take the above statement apart. I do this not out of any wish to triumph over anyone, nor to deny the feeling of hopeless meaninglessness which is part of the common currency of human experience. But I will assert that the feeling is misleading, and the argument behind it is illogical.

First, though, let's acknowledge the feeling. The feeling that the physical world is merely the interaction of mindless forces is a pervasive and very painful aspect of human experience, one that I do not want for a moment to deny or treat lightly. It is, indeed, this very impression that contributes large amounts of the pain of bereavement or of having life-chances cut short. And when you look at the wandering history that led up to the present, you can decide there is no meaning there either (it can seem that way), and nor will there be meaning in the future. 'Vanity, vanity, all is vanity,' said the preacher.

But the argument asserted in the name of atheism goes much further than a feeling. It tries to assert connections or inferences which, even if they are not hard logic in the strictest sense, are nevertheless said to be reasonable. It tries to assert a form of reasoning. It is an argument from properties of the universe to an absence of design, purpose, evil, good, pity, or concern in whatever is the origin or source of the universe. Dawkins is saying that he thinks if there were nothing but pitiless indifference 'at bottom', then the world would come out like it has come out. I think this is wrong. I think, if there were nothing but pitiless indifference at bottom, then we should expect the world to come out showing nothing but pitiless indifference. But it does not. Human beings are part of the world, and human beings do not display mere pitiless indifference all the time. Nor, for that matter, do plenty of other creatures.

Dawkins and other of my intellectual opponents will admit that human beings are capable of pity, and I think we will agree that human beings are capable of good, and ought to restrict and oppose their own tendencies towards evil. The question that remains is, then, whether this aspect of humanity has been spontaneously created from absolutely nothing, from a complete absence of good or pity, or whether it is an expression of a goodness that is independent of humanity and was there all along.

Before we consider humanity, however, let's take a look at the rest of the known universe. To the most brief impression, and to the most thorough investigation, it does not look like a random chaos. When the atheist asserts that this universe 'has precisely the properties we should expect if there is, at bottom, no design', one does begin to wonder. Wouldn't we expect the universe to be a bit less replete with pattern if there is, at bottom, no design?

To handle this, atheism typically advances arguments along the lines of 'anthropic cosmological reasoning', and often it invokes the idea of multiple universes. One idea is, in broad brush, that the fundamental reality that gave or gives rise to all that exists is randomness. Another idea is that the reality that gave or gives rise to all that exists is a collection of abstract principles. I will comment on these. For brevity, the first idea I will call R, the second AP.

Consider R. Perhaps the universe that we are part of is a temporary pattern in a fundamental randomness. Nothing else has shaped the randomness; it is just in the nature of randomness that it can occasionally exhibit patches of structure and order. Patches that have the form of an expanding space-time, for example, with the multiple fields described by quantum gravity theory or some such 'grand unified theory'. Let us admit, this might be so. There is absolutely no way of telling. No matter what signs of pattern we actually find, it will always remain possible to say 'it just happened'. It is the ultimate non-scientific statement, and it can always be made. No evidence can refute it.

Perhaps I am being a little unfair. Perhaps R does have some predictive power, if one adds a small amount of mathematics—just enough to say that even randomness has some properties, namely the fact that in a random situation, structure is rare, and a high degree of structure is rarer than a low degree of structure. If we allow this much, then R begins to say something, because it predicts that the universe around us should be minimally structured. This is a very rough use of the words 'should be', because I am trying not to invoke further mathematical ideas such as probability. Probability is itself a type of structure. But we can admit that if R is so, then we might reasonably infer that the universe has, 'most likely', the least amount of structure that is capable of giving rise to conscious agents such as ourselves. (That last assertion is an example of 'anthropic' reasoning.)

It would indeed be interesting if it turned out that our universe is in some sense minimal. That is, if it is the least complicated that is needed in order to allow long-lived stars, chemistry, Darwinian evolution, brains. It is very hard to know how we could assess that well enough to be able to tell. But it seems to me that the universe has some significant unnecessary 'extras' that suggest it is not minimal. The 'extras' I am referring to are things like deep beauty in the laws of nature, poetry in humans, our sense of justice, and extravagant hope. In a patch of order in a fundamental randomness, would the universe have such deep hidden beauty? And would there be as much tendency to seek beauty and to oppose injustice and to assert hope as is the case in our human lives? Nobody knows. But all the arts, and all the sciences, and all the efforts to assert that life is worth something and we can make a better future together—these all constitute the claim that we can tell what words like *worth* and *better* mean. We can tell. Imperfectly, but such words are not without meaning. As soon as you see that, you have dropped R. You have dropped R because you have said that human brains have a purchase on a type of value and a reality that is not contained in R. This value-laden type of reality is not thrown up by R. It asserts its self-consistent nature quite irrespective of R. R might furnish, by some amazing fluke, the ability to discern value and tell the difference between good and bad, but those things that are good are good quite irrespective of any random chaos that may exist somewhere and somewhen.

Love is not just an alternative to hate; love is *better than* hate. We don't invent such distinctions; we become aware of them. If you think that such assertions have no meaning beyond some sort of genetic pressure or some sort of delusion, then you can revert to R. If you think human beings can whole-heartedly buy into the ultimate meaninglessness of good and bad without losing their poetry and their hope, then you can promote R. But I think anyone who does that has lost a hold on the very rationality that they probably think led them to R in the first place.

To be clear, let me summarize: you and I are allowed to affirm value and meaning in the lives of living creatures such as humans, quite irrespective of their fame or reproductive success or natural talent.

There is no intellectual abdication in so doing. We can be liberated to give full expression to the sense of meaning that we possess, and write all the poetry, music, art, diplomacy, civil rights manifestos, and scientific papers that we can. We do not need to suspect that all this is mere invented amusement that functions to protect us from the 'true' futility of our transient lives in an ultimate meaninglessness. We can quite reasonably suspect that our transient lives take their place in an ultimate meaningfulness.

That 'reasonably suspect' is called *faith*. There is nothing wrong with it. It is not stupid, and it is not arrogant. Rightly handled, it inspires not ignorance or arrogance but hopeful, humble efforts made by ordinary people to lift up their fellows and lift up their hearts to that which renews their hope.

I admit that I have not presented a very carefully argued case above; this piece is more in the way of providing pointers rather than completely annotating the path. There is a risk of presenting a caricature of any position you don't agree with. But I find it hard to take R seriously, because it seems to involve an abdication of the will to seek. It stands at the borders of our understanding, saying 'this far and no further may you explore; beyond there be chaos-dragons.' What does this attitude have to say to anyone who suggests, 'maybe there is more to be discovered'?

Next let's turn to AP. This is the idea that the reality that gave or gives rise to all that exists is *a collection of abstract principles*. This is a very much stronger assertion than R, and presents a very much richer notion of that which is *real without needing to be created or derived from something else*. AP can underpin a form of atheism that one can do business with, because one can discuss what the abstract principles are, and one may find that assertions about meaning and value might be among them. So one might hold AP while rejecting the statement made in the quotation near the start of this chapter.

The reasons why I don't agree with AP are numerous and subtle; I have alluded to some in this book, but I agree that, looked at merely in abstract terms, as an academic puzzle, in isolation from human relations and community, one will never be able to move on from AP. A very powerful argument, for me, is obtained when one moves out of self-centred agonizing and tries to sense what most deeply liberates people to play positive roles. The liberty to receive a sense of being loved and valued and given a positive role to play is one of the greatest human liberties; it causes people to blossom. That this sense can come not only from each other but also from the very ground of being is one of the great treasures of human life, so much so that one should hesitate before one dared to believe it. But it is made available: mostly by showing that what it means in practice can be not just honourable but luminous, and in part by showing that it is intellectually admissible. That it is intellectually admissible can be seen from the intellectually gifted people who have admitted it, and the arguments they have given. In the case of mainstream Christian witness there is a strong tradition of doing this carefully and intelligently. This includes being honest about, and learning from, past failures. I admit, though, that there is also a widespread twitter of well-meaning but feeble argument.

Now I will return to the last sentence of the quotation I started with. The question it naturally invites is: *what properties would you expect to find in the universe if there were, at bottom, some design, or some purpose, or some evil or some good, or something other than pitiless indifference?*

What, exactly, is missing that commentators such as Dennett, Dawkins, and Russell expected to find?

Were they expecting a universe in which pain and tragedy does not happen? Or happens on a much smaller scale? Was it one without earthquakes, perhaps, or without viruses, cancer, and predation? Is it that our lives are too long? Or too short? Or is the problem that our lives are not all the same length, and there is no special protection on the years of childhood, except the one we provide? Or is it that goodness of character does not correlate with ease, happiness, and longevity? Is the problem that we feel deeply the loss of our loved ones? Is it back pain and long-term depression? Or uncontrolled randomness, perhaps? Such a list is not intended to be glib, but to ask the questions in all seriousness. I don't wish to ignore or invalidate the deep problems which pain and injustice confront us with. But when one tries to present examples of how the world could have been better, one also becomes aware of the deep and often inextricable connections between painful things such as bereavement and joyful things such as raising a child. On a planet with a finite surface area, the joy and the work of raising children is only possible because people eventually die. There are also deep connections between pain and empathy, viruses and evolution, uncontrolled openness and freedom. This does not offer a simple picture of a rosy, cosy world, but neither does it permit the equally simplistic analysis commonly offered in atheist rhetoric.

It was not in modern times that people first noticed that life on Earth is painful and difficult. We did not require the help of modern-day activists to state the obvious. The difference between Dawkins's rhetoric in *River Out of Eden* and a poem such as Psalm 90 in the Bible is the tone of voice: the first is strident, commanding, contemptuous of meaning; the second is mournful, yearning, hoping for meaning.

Now, in venturing to ask the above questions, and in making the point that there are connections between what makes life painful and what makes life fulfilling, I do not for a moment intend to suggest that arbitrary and extreme suffering can thus be justified. I certainly hate, with a sustained hatred, the notion that this or that random affliction has been caused in order to achieve some good, or can be justified by what good may result from it. Rather, in the above I am simply pondering, briefly, the truth that has been much more deeply explored in the arts. This is the truth that our life of finite resources and limited powers *is*, simply and directly, who we are, and this is our limitation, and this is also our glory.

If one dares to dip one's intellectual toes into the question 'what would a world look like if it were the creation of a profoundly good and able supporter?' then one quickly realizes that a world that was painless and incapable of improvement could, of logical necessity, have no opportunity for courage and no one in it with meaningful work to do.[66] It is even doubtful whether there could be any sense of self or freedom unless the world itself is radically free to go wrong as well as go right. But this is not to say that the universe as we know it could be said to be 'the best of all possible worlds'. Rather, one begins to sense merely the emptiness of that phrase, and one grows weary of those who think they can pronounce judgement on the whole scheme of the world. The statement which this chapter takes issue with begins to seem less and less brave and honest, and more and more like a kind of posturing.

Who knows what we should expect if there were, at bottom, no design, no purpose, no evil, no good, nothing but pitiless indifference? None of us. No human being, whether alive now or in the past, has ever been in a position to pronounce on a question like that. We lack both data and intellectual capacity. We experience both the pitiless impersonal forces of nature, and also all that goes to make life valuable. The reasonable position is not to buy into the rhetoric which chooses the former over the latter as the deeper or the prior truth. The reasonable position is to admit the sheer implausibility of the idea that those who pontificate on the absence of either good or its opposite in the basic scheme of the cosmos even remotely know what they are talking about. But rather than passing judgement on the world in pompous style, another option is available to us. Rather than trying to stand in overall judgement we can express, by our actions, a sense of solidarity with the project of the world. We can express our willingness to co-work with what good the world can produce. We can commit ourselves to whatever good we can sense, and insofar as life on Earth has been made better, it is because people have been willing to do this. And that very fact is a fundamental good of the nature of the world.

During the course of my writing this book, a friend put it to me that compassion and empathy are not conspicuous in the natural world, beyond their necessity for survival purposes. I think that, in the case of the human community, this is untrue, and in the case of our mammalian cousins it is unclear, because every social group has some sort of empathy going on. But, whether or not this sort of observation is correct, what I want to know is the answer to the question, are these qualities ways of getting at truth, like, '2 + 2 = 4', or are they meaningless beyond their survival value? That is, does the admonition 'it is good to trust others and to build a shared and equal future' express a beautiful truth, like Fermat's last theorem, that would be true whether or not any human ever lived, or is it just a convenient way to speak, or perhaps a delusion helplessly entertained by gene-controlled robots? Is it a way of thinking that came about *only* because human genes will proliferate if we adopt this sort of attitude?

It may indeed be true that without ethics humans would die out, but as a matter of fact that is not uppermost in the minds of most people who choose to live ethically. What we experience is a sense that others and ourselves have innate value, and we begin to develop a sense that another's sorrow and joy matters to us, no matter what may happen with regard to mere survival of the species. And we go further. When we watch the behaviour of the shoebill stork (which rejects one offspring in favour of another) we feel uneasy; the sense of connection that the parent bird lacks is felt by ourselves.

What I am doing is taking this experience seriously. I am saying that I experience a sense that in ethics and morality, and in solidarity and loving concern, we are gradually discovering a domain of truth with as great a claim to objectivity as anything else humans are capable of grasping, and I have checked that this sense has good intellectual credentials. It is not a provable item, but it is one which can be lived by. Following this thought through, I find that, just as any natural process cannot help but respect basic facts of arithmetic, and just as biology cannot help but work with the fact that round shapes are more efficient than squares, so also, sufficiently aware things like embodied persons cannot help but become aware of moral truth. So there is, 'at bottom', or 'at top', or wherever, this type of truth, as well as the truths of mathematics and of art. My contention is that this way of understanding our situation has as great a claim to intellectual credentials and scientific credibility as any other of which I am aware.

Before concluding, it will be well to reiterate one point. I have already asserted that it is objectionable to try to justify all forms of suffering (though there are some forms that are justifiable). The arbitrary traumas that come about owing to the uncontrolled nature of things are not to be justified but recognized as *outside the categories of order*. Arbitrary and unjust loss and pain has been termed *moral disorder*. Moral disorder flattens our reasoning powers; it refuses to budge. It cannot be wrestled into making sense; sense is precisely what it does not make. So we have to give up trying to make sense of it. The correct response is to name it as as shitty as it truly is, but to decide that our reaction will be one of compassion (whether towards ourselves or others), creative intelligence, and hope. We are called upon to meet such situations with what empathy and help we can bring to bear.

To return to the words of Dawkins's statement again: what if there were, 'at bottom', some good? Then we would expect to find that the universe is, at bottom, good. But that is precisely what we do find. Even after acknowledging the imperfection and unfairness, one should be very hesitant before buying the idea that the world is neutral, as if we could, with moral integrity, choose not to commit ourselves to furthering a purpose in it which we did not invent but recognized and responded to. And the situation is stronger still. It is not just inaccurate, but also morally repugnant to denounce the whole natural order as basically bad or indifferent, because such a statement implies either that it ought to be stopped completely, or that it matters nothing one way or the other whether it continues or not; the logical implication is that it would be correct or morally neutral to institute a campaign of universal destruction or relentless apathy about the future of the world, and this is wrong. It is about as wrong as can be. It is the way of madness. It would be the action of some sort of insane or morally bankrupt ideologue.

I say the world is immensely precious; I utterly refute the notion that it is not.

In many respects the world is also fragile, and this means I also oppose the practice of announcing the world to be value-free, because, in addition to being untruthful, that practice reduces the sense of hesitation that one might otherwise feel before making a species go extinct, or before reducing a mountain to rubble.[41]

However, I will agree that the universe, or to be precise, the patch of it where we live, is not as good as we can help to make it. It is a work in progress, and the solution, and also the biggest problem, is largely ourselves.

It is important to acknowledge that the world is good enough to merit our commitment to working with it, not abandoning it. Apathy and suicide are neither correct nor adequate responses to the universe. But this does not make the problem of pain go away. Part of that problem is the fact that if there is some 'higher' or 'deeper' reality that can be held responsible for originating the universe—if that theistic way of thinking makes sense—then we start to feel that we or others have been hard done by. We feel caught: we have to be grateful, we suppose, for having a chance to exist at all, but the existence we have is not a fair one. Are we allowed to complain about this? According to biblical authors, such complaints are understandable and are not to be ignored. But we won't get any response other than absolute reality. The universe is as it is: wonderful, difficult, and not fair, and our complaints won't change that. The only response that comes from absolute reality is to present us again with the actual world itself, and our role in it: to let us know that we have a role, and help is at hand to establish justice for others, not joy for ourselves (but the latter will come from the former; 'seek first his kingdom and his righteousness, and all these things shall be yours as well,' Matthew 6:33). Any complaint we raise against injustice is only real, as opposed to mere vapour and posturing, if we act as well as speak—if we join the struggle to relieve distress and reduce affliction—and as soon as we do that then we are back on track. We are doing business with the absolute reality from which our lives have come and to which they can legitimately be offered. Also, according to the Christian message, there is more that can be said. According to us, the 'management' is not just offering resources (perseverance, truthfulness, scientific insight, etc.) but is also engaged in a more intimate way. The buck stops with one who shares the pain of the cosmos.

What would we expect if there were some purpose, something other than pitiless indifference? We would expect, I suggest, that agents capable of sensing purpose (and that means conscious agents) would sense that there is purpose. And that is precisely what most of us do sense. Our lives have purpose. The purpose is, approximately, to realize a more and more complete expression of goodness, beauty, and truth, in the context of imperfection, improvisation, and shared resources. And to crack a few jokes along the way.

Furthermore, we sense more than purpose. We sense the transcendent. It comes at us again and again, in mathematics, in the marvels of the natural world (whether directly experienced or in scientific analysis), and in a glimmer that sometimes lights up our moments of shared hope. Sometimes it breaks in to ordinary moments of humour, and even grief, when it is a healthy grief, can become a modified darkness when we feel it to be shared and comprehended by that which holds us.

I recently read the novelization by Michael Punke of the life of Hugh Glass, the nineteenth-century American fur trapper who was mauled by a bear and abandoned, but rescued himself through determination and survival skills. His story is based on some small snippets of historical information and is probably overlaid with legend from an early stage, but it set me thinking more generally about the way ordinary people have, the world over, and throughout history, put their life together out of a hotchpotch of things and ideas available to them, and made something compromised and imperfect, but worthwhile. The natural world that has so much indifference to us nevertheless inspires in us intimations of a depth and a grandeur that is, at times, utterly liberating. We are drawn out of ourselves and find at once an inexhaustible array of things to learn and appreciate, and a sort of respect for who we are, puny though we may be.

It is the impersonal forces of nature that exhibit pitiless indifference. Of course they do. That is their nature. But what did we expect? Some sort of reward system for good behaviour? That would make our lives utterly banal. Instead we have meaning and the invitation to join in. And if we explore, tentatively, in the direction of pity—not in the impersonal forces of nature, but in what we encounter in one another and, notably, in prayer too—then we do, yes, discover something that is not pitiless indifference, but quite the opposite. We discover a sense of solidarity, a sense that our difficulties are fairly assessed and our pain is shared, and we discover a great encouragement to live creatively.

I have argued, then, that, notwithstanding its difficulty, the universe is neither bad nor neutral but good. I have admitted that I don't have the intellectual resources to assess and announce that conclusion with any sort of intellectual authority; I have merely asserted that it is a reasonable position, and arguably more reasonable than others, and in any case my whole being is drawn to it and I have voted for it. I have also argued that the natural world appears to have a degree of depth and richness that exceeds what is necessary for the mere existence of conscious agents, though we cannot be sure of this. I have argued that there is every evidence of purpose being realized as we take up our responsibility to care for each other and the rest of life on Earth, and as other beings—animals, plants, rocks—enact what good they can. I have argued that the natural world is also the vehicle of transformative and transcendent experiences. The statement in the quotation from *River Out of Eden* with which I started is almost completely wrong. The universe that we observe has a mix of properties, and included among them are some that we would expect if there is, at bottom, design, purpose, and good.

Performance

'Friends, citizens, intelligentlemen,
come all who are eager,
receive the elixir of life.
It is another comforter
to bide with us forever:
logical curiosity.'

I saw you fashioning your case,
pirouetting your passion,
lyrical with logic which indeed
awakens the sense of beauty—
a universal sense—
but something was awry,
something hard to pin down,
a lurking smirking.

I followed along on the sweep
of your wave,
and, good disciple, I
finished by trying to
regard you with
logical curiosity.
I began to attend to you
with that alone: pure elixir.
And a sense of horror crept over me,
not that you had died, but that I had.

At the end of the lecture,
someone asked you how you could be so sure.
For some reason this elicited
not a reply but a rebuff,
a strutting ridicule.
But I felt the question was honest enough.
It was your confidence made me hesitate.

PART III
BREATHING

18

Silence

Try the following as a spiritual exercise.

Take an ordinary inanimate physical object, such as a stone, a cup, a thread, a bicycle, or a pen, a computer, a cloud, or a crumb, and sit yourself down in front of it and look at it.

Don't say anything, and don't think anything either. Just enjoy. Enjoy the physical object for what it is. Don't go out and grasp it with your mind. Let it come to you. Look intently. See. Receive. Don't completely suppress all your analytical thoughts—the thoughts about physical and chemical composition, and history and usefulness and uselessness—but don't marshal them, just let them ebb away. Today they quieten in order to allow a more direct and simple appreciation of this ordinary, fascinating, wonderful physical object.

Don't assess. Don't appropriate the object. Merely receive whatever it gives freely. Don't interpret it, don't reject it, don't own it, don't disown it. Be with it.

Smile with it. Silently laugh with it. Or be in sadness with it, and be bereft with it.

Do not speak to it—it cannot receive speech.

If it is within reach, put out your hand to feel it, and to weigh it. Let it resist you. Let it be.

Let it be, let it be, let it be, let it be.

Let yourself be.

The above is called a spiritual exercise because it is an action to undertake, an exercise like going on a twenty-minute jog, or going without a phone for a day: in order to get the benefit, one does not analyse it, one just does it. It is included in this book because it makes space and a pause. The aim is twofold. First, it is to provide an opportunity to experience and acknowledge the wonderful, unspeakable, mysterious marvel of any ordinary object. Second, it is to take a moment to deliberately switch off the endless attempt to appropriate, analyse, and interpret our surroundings, and instead let them impinge on us in another way. It is good sometimes merely to be one part of the world quietly allowing another part of the world to make its presence felt.

This exercise is an example of, or is closely related to, the practice called mindfulness. The above was written several years before the idea of mindfulness became fashionable, as it has done recently. I included it for the reasons just given. Mindfulness is not exactly the same as prayer, but they have in common that their aim is truthfulness and an effort to make oneself and one's opinions not the centre of one's world. Prayer is not something I will try to discuss further here, but I will mention merely that prayer is more about silence than it is about talking.

Universal Acid

There is an acid, they say,
working at the bars of thought,
whose disinterested action dissolves unstoppably.
Once spilt it drills down,
like the blood of the creature in *Alien*,
and reconfigures everything.
In powerless fascination we watch it
piercing all our decks and threatening
to breach the hull. We rush to catch it
in old crucibles which dissolve at its touch.
Titanium arguments ooze and bubble.
Porcelain logic softens and sags.
But just before the acid dissolved us
we found, to our strange delight,
that we could catch it in our bare hands.
We stand or kneel and watch it
pooling there and wonder who we are,
what kind of creature made of such blood?

19

The Human Community

> The worldwide community of people is considered. Our evolutionary story is briefly sketched, including early human expression such as cave art. Our aesthetic ability, reasoning ability, and moral ability are considered. All of these are end products of physical processes; all are nonetheless genuine for that. The same goes for our religious sense, which is the aptitude for discerning meaning. I look at some of the variety of forms of religious expression, and comment on Christian history.

The long process by which different sorts of living things grew and reproduced on planet Earth has been the process of our birth. A form of chemical replication probably happened in some sort of molecular soup in tidal pools or ocean vents billions of years ago. Somehow, simple cell walls and other wonderful structures such as DNA molecules were 'discovered' by a process of constrained and stimulated trial and error: constrained by physical and mathematical conditions, stimulated by the global heat-engine of sun, Earth, and space and by the universal exploratory capacity that is deeply woven into the physical world.

Things like muscles and a skeleton proved to be useful accessories for some creatures; bark and leaves were useful to others. Light detectors of one form or another were very useful, and were discovered several times. Chemical receptors furnished an equally useful sense of smell and taste; pressure sensors gave a sense of touch, and delicate structures gave further senses such as hearing and balance. Thus creatures became receptive to a wider and wider range of aspects of the environment which shaped them, and on which they depended.

Eventually, some hundred thousand years ago there were hominids with a physical anatomy essentially the same as that of modern humans, except in the detailed structure of the brain. They lived alongside one another for tens of thousands of years, during which their range of skills and experiences slowly widened. They used fire for cooking and

Science and Humanity. Andrew Steane, Oxford University Press (2018).
© Andrew Steane. DOI: 10.1093/oso/9780198824589.001.0001

clay-hardening, for example (and there are hints that that particular skill goes back to much earlier times).

By this steady but slow process, eventually new capabilities were attained, and their comparatively rapid appearance suggests that some sort of threshold was passed, or a feedback mechanism was begun, which allowed more rapid development of the brain. Perhaps forty thousand or fifty thousand years ago, people developed a rich variety of complex behaviours, including artworks painted on rock walls, clay statues, clothing decorated with beads, sharp arrow points, a variety of flint tools, and the like. These early people had many interesting conversations as they thought about their social arrangements, how to raise their children, how to manufacture more and better tools, and as they admired each other's artworks and prowess at hunting or cooking or tool-making, or singing, running, throwing, wrestling, and dancing.

From the complexity of their artefacts, we can deduce the complexity of their thought processes, and of their language. These were thoughtful mammals who almost certainly had the further, more subtle, senses that we detect in ourselves: senses such as the aesthetic sense, and the moral sense, and the religious sense.

It is possible for us to stand before some aspects of the natural world and feel, both strongly and deeply, the sense of beauty. It is possible to describe this experience as if it is only about human psychology, as if beauty were some sort of convenient delusion by which some people make themselves happy some of the time. But that way of talking about it, as if beauty were merely a human invention or a figment of imagination, is not true to the experience itself. Beauty confronts us, addresses us, moulds us, calls us. The clarinet concerto by Mozart is not just different from the clarinet concertos of Weber; it is a more deep, true, and beautiful realization of what a clarinet concerto can be. The idea that the aesthetic sense is a sense of something objective has strong philosophical credentials, and I would like to encourage you to take it seriously. It implies that we have more than just the physical senses of sight, sound, taste, touch, balance, temperature, and hearing, but also the sense of beauty. The painters of the caves some thirty to forty thousand years ago had it too.

We have a reasoning ability which involves a further sort of sense: an appreciation of concepts such as number and logic. If, for example, we are setting out food for the family, then we can readily match the number of apples to the number of children without having to have them all present. Or if two further friends arrive, then we know to add two more apples to the pile. Equally, we perform elementary logic all the time. For example, if dry wood generally burns better than wet wood, and this grey log is drier than that green one, then we are able to 'put two and two together' and deduce that the grey log is a better bet for getting a fire going.

We also have a moral sense. This is shaped by a coming together of an innate potential to develop moral receptiveness that is part of the birth condition of every healthy baby, and a wealth of taught ideas and behaviours by which we shape the way our children grow. Those taught ideas originated in the human race by a process of inspiration, and a shared effort to discern what is best. People repeatedly and courageously did not allow the bullies and the charlatans to win the hearts of the next generation, and thus our lives and our understanding of goodness were enriched.

Here I must briefly point out one mistake that is commonly made nowadays. This is the argument you sometimes hear, that because the human moral sense came out of evolutionary development through natural selection, followed by cultural development, it follows that the moral sense is arbitrary. Actually, that is not an argument. It is a non sequitur: the conclusion does not follow from the premise. Through the gradual and tremendously creative process of evolutionary development, what happened was that brains became more and more sophisticated in their model-making capabilities, and they also became capable of sensing and appreciating richer and deeper aspects of what is real and makes sense. Any sufficiently sophisticated, model-making, self-conscious entity, capable of appreciating that other agents also have interests and consciousness, is also capable of acquiring moral knowledge, and this is what happened in the human race on planet Earth. By 'moral knowledge' I mean, for example, basic principles such as that one person ought not to subject another to slavery or other sorts of brutal treatment. This principle is not an arbitrary invention or a mere convenience; it is a precious truth that we have, together, discovered, and no amount of philosophizing will undiscover it. We will one day find a better approximation to the law of gravity, but we will not one day discover that apples released near to planet Earth do not fall towards Earth. Similarly, we will one day discover a more nuanced understanding of human relations, but we will not discover that slavery is OK after all. The moral sense that enables us to know this is a precious perceptive ability supported by our cognitive apparatus, rather as our aesthetic sense is.

The moral sense only comes into play once a child is old enough to understand such things, of course, but as soon as they are old enough, it is perfectly possible to tell a youngster that he *ought not* to torture the cat. In other words, the child has some innate sense of what words like 'ought' and 'right' and 'wrong' refer to. An abused or misled child will, sadly, develop some wrong ideas about what is good and what is not, but the moral sense is here misshapen, not absent altogether.

Finally, we have a religious sense. It is the sense that we are part of something larger than ourselves, or called upon, and the sense that what surrounds and holds us is a carrier of meaning.

Here we must tread carefully, because religion is so easily and so widely abused. In writing the final chapters of this book I am in some difficulty because the idea I am trying to share is one I have only got a partial grasp of myself, and in any case it can't be shared through written words alone. But there is a sort of music about our lives which we can catch if we listen carefully, and which we can join in with when we are ready, and it is a way of seeing and living. It is a way of seeing others and yourself; it is a way of understanding that, whatever your or my situation—and I really mean *whatever*, no matter how constrained or free, no matter how easy or painful—you and I have a meaningful role to play and valued people to be, as part of a community on which we depend utterly, and which we also support by our service, and in which we are free to exercise *every* aspect of our humanity, including the aspect commonly called religious, the deep part which would be set free if only you could find an expression of it that you could give and receive in good conscience.

It is a deep part, and an innate part. The innate apparatus with which people are born includes, as well as other things, an aptitude for language, and an aptitude for detecting and sharing meaning and purpose. Both of these aptitudes are amply demonstrated by psychological studies, and since they are such deeply ingrained aspects of the structure of our brains, we can assume they were present also for early people some fifty thousand or so years ago. The ingrained religious sense is currently being studied in various ways, and because this is a controversial area, its interpretation is not universally agreed. However, it seems broadly established that it comes very naturally to human children to consider that the natural world is purposeful, and that where immediate knowledge and experience finishes, there is more reality to be sought, beyond what we immediately know. There is something or someone 'out there' or 'around' and also to some extent 'in here', in our innermost thoughts. This can become quickly overlaid with storytelling, not all of it making sense of course, but the point is, this way of thinking is not something foisted on five-year-olds by adults with controlling agendas, but an aspect of human nature that comes naturally. This doesn't show whether such ways of thinking are productive and insightful, or unproductive and foolish, of course, but it does suggest that one idea commonly assumed in the past by Western anthropologists, and often assumed by modern people, is in fact wrong. This is the idea that religious thinking in the human race started out in animism and polytheism and slowly developed towards monotheism and atheism. In fact religious thinking started out in the human race because humans possess a sense for objective religious truth, just as they do for aesthetic truth and logical implication and moral truth.

Yes, I insist, religious truth, not religious invention. We each possess the sense, the hint, that muggins is not just making a largely farcical journey into an amoral future, but muggins is making a largely meaningful journey into a future in which our behaviour is cared about. Some say it is only cared about by ourselves and other animals. But people have very often felt that their actions were also cared about by something or someone else, some sort of hard-to-describe holder of value, which we aspire to live up to. Don't listen to those atheist doctors coming running over the fields in their long coats, hastening to reassure themselves that all this is merely an offshoot of some other agenda that gave an evolutionary advantage. What the human species has received from our Darwinian evolutionary upbringing is a vast amount of trial-and-error-honed sense for what is real and what is not. And the sense of a moulding influence on us—one met with at times of exquisite beauty and at moments of hope in grief, one also singing in the songs of the birds and waving in the leaves of the trees, one that is truly deeper and greater than ourselves, and which really could deserve our allegiance—is a sense of reality not unreality.

I am here trying to do a fair job of reporting human experience, and I happily acknowledge that religious experience throughout the world does not everywhere express itself in straightforwardly theistic terms. I also happily acknowledge that a principled atheism deserves the respect due to all well-lived and well-argued positions. But I don't think it is in the least convincing or humane to dismiss all religious experience as not even worthy of recognition in the category of evidence, as just so much fiction springing up from the fertile and furtive human imagination. Atheism in practice can represent sheer laziness of thought just as much as any other view can; it is not the bastion of rationality it supposes itself to be.

Here I mean by the term 'religious experience' not just mystical experience, but also the making sense of the everyday, and the carefully constructed sequences of thought that are captured in revered writings, and the myriad simple acts of gratitude and hope that constitute large parts of religious expression.

In the various Taoist schools, the theme is wisdom and wise action, but it is an important component that wisdom is not something that can be fully broken down and examined from a distance, and some at least of what theism ascribes to God is 'read' as part of the nature of the physical world, in, very roughly speaking, a pantheistic sense. The various Buddhist traditions are also cautious with language and stress the refusal to accommodate cravings, including the craving to have simple answers to existential questions. Thus Buddhism does not officially endorse the notion of a loving reality which affirms each person as a person, and indeed it is highly paradoxical because it appears to deny the very notion that there is in any of us a self worth loving, yet it succeeds nonetheless in encouraging the expression of calm respect. This paradox has a resonance with the thoroughly Christian idea that to save one's life one must lose it. That is, a fundamental act of giving up of the self is the gate through which we each have to pass. The difference between a Buddhist and a Christian* understanding of this is that we learn that we give the gift of ourselves to One who lovingly gives us back.

Buddhism is more correctly termed agnostic than atheistic, and this hesitation in pronouncing judgement is an attractive feature. However, if one looks at human behaviour, it is noteworthy that people do find it natural to act in distinctly theistic ways. The practice of Buddhism has very often involved a reaching for, or a sense of, an affirmation of each person for who they are, coming ultimately from beyond the human

---

* I occasionally use the word *Christian* in this book, but I do so with a sense of hesitation. I find that the people who are quickest or loudest about selling their views under this title often adopt attitudes that are quite unlike the way Jesus of Nazareth went about things.

community which is its immediate locus of expression. This can be seen by the way Gautama himself has, with temples and statues, been revered as a god-like being, despite his express wishes and teaching.

I mention these brief observations on non-straightforwardly theistic religion in the interests of getting a fair appraisal of human experience in the round, including all the ways people have found to help each other live more wisely and generously than they might otherwise do. None of the religious traditions have been uniformly generous or wise; all the major ones have some value to offer. I don't wish to imply that all religious ideas have equal intellectual, moral, and spiritual credentials; that would be very unlikely to be true and I think it is not true. For example, one can understand the urge towards a form of pantheism, but after careful consideration I don't think it stands up to philosophical scrutiny, and it is overturned by more telling evidence from other historic and present-day experience. The world is too ambiguous to be revered in all respects—we must abolish the Ebola virus not accept it—and pantheism should give way to the evidence of what happened to the ancient Hebrews and what happened to Jesus of Nazareth and what duty this places on us. However, the idea of reverence in our dealings with the natural world is both valuable and powerfully creative when exercised judiciously, and the idea of God becoming incarnate in the whole of what is can be received as a statement of what the universe might be gesturing towards—a kind of existential blossoming in future time, or out of time, rather than a present reality.

In Chapter 17 I discussed the sense in which the natural world need not be interpreted as a sign of indifference, even though many of its processes are indifferent to us. Here I will, tentatively, take this a little further by presenting some thoughts that the *gospel* accounts suggest to me.

The physical world of ordinary, fragile stuff commands our utmost attention not only because it embodies ourselves along with the rest of the biosphere, but also because it is the medium chosen by God for God's self-expression. However, I have already stated more than once that the world is ambiguous, so in what sense is God expressed in it, or how can one possibly say such a thing, in view of the senseless pain of the world? The attitude which I think one meets in the provocative public spokesman described in the gospels is to decide to be on the lookout for what good can be accomplished, in any situation, and join in with that good. That plain-spoken northerner doesn't inspect the gift of physical existence with a view to determining whether or not the giver is loving; he starts out from complete trust that the giver *is* loving and interprets the gift accordingly. The domain of generous, creative relationships, which he referred to as the 'kingdom of God,'* is not somewhere else, it is right among us here and now, if only we understand it and will it so. To speak of '$\vartheta\epsilon o\varsigma$' ('God') is, for one who knows what they are talking about, simply to be alive and responsive, alert to the fact that something precious is being accomplished through the world we are directly aware of and part of. In the accounts of his life, when Jesus encounters the indifference of the processes of nature, he interprets according to his view of God. A difficulty such as congenital blindness is an opportunity to live what good can be shown in a life constrained by that difficulty, and also an opportunity for us to relieve the burden if we can (John chapter 9). In the case of a tragic fall of a tower, he resists the idea that it was either deserved or undeserved, using it simply as a reminder of mortality (Luke chapter 13), but in the case of the life-giving rain which falls indifferently on the just and unjust alike, he is inclined to adopt it as an illustration of generosity (Matthew chapter 5). Keenly attentive to what goes on in the tangible physical world—the lily, the grass, the seed, the sheep—he mines it for parables and illustrations of the nature of love.

Another test which religious and irreligious traditions have to undergo is the test of how well they championed the poor in practice. This includes how quickly and decisively they moved away from forms of society based on hegemony in social relations, in which slavery, or something roughly equivalent to it, persisted, and how well they now receive and enact fair and liberating forms of government.

Any assessment of human history must grapple with the remarkable impact of the ancient Greeks, the ancient Hebrews, and the early Christian community which fused the inheritance of those two. These stand out by any objective assessment, in their influence in comparison to their numbers and political power, in their championing of notions of universal justice ahead of their time, and in the courage and common purpose they showed.

Jewish culture retains a striking impact in the arts, science, commerce, governance, and learning right down to the present day, and

---

* The phrase, as Jesus employed it, might equally well be translated 'the realm of God' or 'the commonwealth of God'.

an extraordinary resilience. The events of first-century Palestine also deserve profound attention. I think that that was a time and place where our local part of the universe, our humble home, saw an exquisite blossoming of the moral and spiritual principles that infuse life at its fullest. This did not take place in a paradise or a myth but in the messy backwater of a small country under military occupation. In that setting we are shown both what it is to be human, fully and without evasion, and also what is at the heart of being: a seed growing in secret, a treasure worth everything, a seeking and being sought, free forgiveness, open-heartedness, a bearing of pain, a total giving, an astonishing renewal.

Such insights do not become outmoded or dated any more than the principles of arithmetic do, but we do make progress in grasping them more fully and there is a never-ending need for fresh and imaginative expression.*

Christian witness has been both beautifully impressive and also thoroughly compromised in its long history. It has been compromised especially by anti-semitism, by flirting with power and colonialism, by slowness to allow full freedom of speech and citizen rights to atheism, and by overly negative or fearful attitudes to sexuality. Nevertheless, for all its failings, Christian witness has proved itself deeply reformable and capable of expressing a positive contribution in a rich variety of cultural settings across the world. That same witness has also offered great artistic expression (I mean the gifts of people like Dante and J. S. Bach, Dostoyevsky and Gaudi, not the strand of modern-day pop culture that calls itself 'Christian'). In matters of sex there is also a very great deal on the positive side of the account, in the support given to the freely chosen loving commitment of ordinary people to one another (not the dubious power games to which the wealthy debased their notion of 'marriage'), and in maintaining structures designed to support the sort of long-term commitment through which both children and adults are given a security which deeply liberates them.

I omitted misogyny from the list above because, although Christian bodies have participated in misogyny, it may be unfair to single them

---

* There is also an ongoing crisis in all religion that has not yet embraced the freedoms won by the Enlightenment, especially the generous instinct of support for honest free enquiry. In the Christian case there remains a large 'mole' on the overall body, in the form of a belief system in which trust in God is replaced, broadly speaking, with contorted worries, and everyday effort for service of others is replaced by effort to grow God's doctrinally correct 'in-group' who will make it to an afterlife.

out for this, and many laws which look regressive now were progressive in their time. Women have often been drawn to, and felt validated by, what goes on in churches.* Perhaps this is because, even though church organizations lost sight of the leadership potential of women for a long time, those same bodies have held people to high standards of behaviour, and have often acted as a focus for activities where women were more free than they were in other parts of their lives. None of this makes up for the viciously misogynistic rhetoric which has appeared at various times, because moral failure can never be compensated for in that way. Rather, one recognizes failure and changes one's thinking accordingly. The issue for us is, are *we* willing and able to be fair in our assessments, and can we recognize good as well as bad in places where it might be inconvenient to our assumptions?

Religious leaders in Christian monastic and church communities have sometimes descended into aggression, but more often they have acted to question the warlike impulses of kings and princes in the past, and secular governments in the present. Furthermore, to a very significant extent, committed Christian discussion and effort has been the place which nurtured and gave to the world the notion of universal human rights, the practices of empirical science, secular democracy, universal education, and universal welfare. I say 'to a very significant extent'; this statement is not intended to deny the role of other points of view, but it is a plea that spades should be called spades and history be fairly assessed not rewritten. Magna Carta, for example, was conceived out of a coming together of secular and religious opinion; it was drafted by a bishop, opposed by a pope, and finally settled following a period of negotiation between crown, church and, barons. It is quite wrong to say (as one commonly hears today) that religious concerns uniformly gave a false legitimacy to the concept of royal rule. In fact they were sometimes used to do that, but more often religiously motivated bodies acted as the chief opposition to royal power. The journey to the modern concept of secular democracy began not only with the ancient Greeks, but also with the insistence of the ancient Jewish people on self-rule, and with the followers of Jesus of Nazareth who refused to regard the

---

* 'Despite common misconception, Christianity is similarly generous and embracing of diversity at its core, being based on the example of a man who empowered women by welcoming them as his disciples to be taught alongside men, by ignoring ritual purity laws which discriminated against women, by standing up for prostitutes and the rights of women when it came to divorce.' Jemima Thackray, *Telegraph*, 13 February 2014.

Roman civic authorities as sacred, and were executed for this. The Protestant Reformation and the English Civil War were in large part about insisting that what is sacred or non-negotiable is to do with greater freedom for people to work out their lives by conscience and negotiation; the sacred is *not* to do with centralized power. What Jesus of Nazareth gave his life to declare is the startling fact that not even God is about centralized power.

I have entitled this chapter 'The Human Community', not 'The Human Individual', because we can only make sense of our lives as a community. The expression of human life spreads across a range of psychological types and physical abilities, and it falls to some people to have to manage at the extremes of these ranges. The psychological range extends from what may loosely be called autistic at one end to schizophrenic at the other. I am using these terms loosely to refer to, at one extreme, the mind that focuses on one thing at great length, and at the other to the mind that jumps haphazardly from one idea to another. The world, especially the biological part of it, is not fully controlled, and the result is both our glory and our difficulty: all the range of ways of being human arise. A mildly autistic way of thinking can be very fruitful in many types of problem-solving, and a mildly schizophrenic way of thinking can be very fruitful in many types of creativity. But, further out, there are ways of being human that are innately precious, every bit as valuable as any other way, but which are very difficult to live out. It is the job of all of us to even out the life opportunities, as far as is possible: to help the loners to know that they are not alone, and to help the unsettled to feel that there is something steady and dependable to which they can return.

Further out still are the lives which cannot be lived, except for a short time. Those which, because of genetic problems or serious illness at birth, will only survive for days or weeks or a few years. These too are part of our community, and our community only makes sense if we recognize them too, seeing our connection to them, and theirs to us.

20

Encounter

> Science accounts for what is going on; religion encounters what is going on.

The proper practice of science is *part of*, not an alternative to, a humane philosophy of any kind, and this includes its being part of, not an alternative to, a theistic way of life and outlook on the world. In the latter way of life there are not two things, 'science' and 'religion', but one thing, *faith expressing itself through love*, and this one thing expresses itself in laborious, patient, joyful application of reason to the study of the natural world, and also in laborious, patient, joyful activity of other kinds, including prayer and shared rituals celebrating the things we value highly and receive daily.

It has been a theme of this book that, in addition to photons and electrons, quarks and gluons, and things like that, there are other realities we have to reckon with: things like logic and mathematics, aesthetics, and also moral principles. For example, the principle that you shouldn't squash, bribe, degrade, impoverish, do violence to, or in any way objectify other people, but rather nourish, encourage, respect, feed, protect, and pay attention to them. (That doesn't rule out opposing whatever violent impulses they may show, but such dilemmas themselves call for wisdom and restraint).

Many people are alert to the idea that when a person is before us, something infinitely more precious and profound is before us than a mere animated pile of molecules. We are in the company of a *person*, someone who calls for our attention as one person to another. Anyone who can sense this, who can genuinely receive and act on it, is in possession of a connection to, or sense of, a good that is the source of all goodness. Atheism is the opinion, often connected to a form of reductionism, that this good is fundamentally impersonal. Theism is the opinion, grounded in experience and accompanied by reason, that impersonal categories are inadequate to capture the nature of this good.

Science and Humanity. Andrew Steane, Oxford University Press (2018).
© Andrew Steane. DOI: 10.1093/oso/9780198824589.001.0001

This good is the single ever-present *other* that people sense more fully when they give themselves a chance to do so, by mindfulness, by frankly asking for help, and by pondering the lives of people or communities who demonstrated humanity at its best, under the constraints of each time and place. Such people express a sense of indebtedness, of having been the channel of something that came from beyond them—from others who went before them, perhaps, or somehow from the world at large and the very way things are. This lovely mystery to which we are indebted is the good that is the love that creates. This is the love that transfigures analysis. Without that love all scientific analysis is ultimately doomed to objectifying humankind en masse and thus destroying us. With that love, on the other hand, science is liberated to serve its true purpose, which is to make us co-creators of a better future.

Faith is willingness to recognize and trust this. Such faith is completely comfortable with, good at, and faithful to the spirit and demands of science, in every area of investigation, to the highest standard. This does not mean introducing alternative accounts of natural phenomena that don't pass muster with the honest and well-argued opinion of the general scientific community. It means joining with that community in an intelligent and good-hearted effort to find things out, while remaining alert to the possible abuse of science. Examples of such abuse include: environmental devastation, facilitating unjust trading practices, the claim that science is corrosive to a rightly understood theism, the idea that human beings can correctly be regarded as objects.

21

The Human Being

The nature of human *being* is considered. To be human is to be a part of a network of connected people who bear with one another, and teach, support, receive, and give one another. Our life comes both from below and from above: that is to say, both from the physical structures of the world, and from the shaping influence which moulds what those structures can express. This is especially true of the way we see ourselves and each other. Humans both build on existing resources, and also receive creative inspiration. This inspiration is not able to be fully captured in impersonal language. We find our life in full by being willing to admit this.

Imagine

Imagine someone told you there was a photocopier in the corridor.
Now imagine someone told you there was a tiger in the corridor.
Would you feel different? But nothing has changed,
except that you have seen your situation differently.

Imagine someone told you the floor, which is the floor,
is a myriad of electric pulses and knots of
charged effervescence in intricate interlinked dance.

Imagine someone told you the sky, which is the sky,
is a boisterous tumbling of dots and commas
endlessly seeking space.

Imagine your arm was a force field, your hand a
nexus of inward–outward sense and shaping touch.
Imagine your brain was a jungle of living fibres
silently approaching and receding from each other's kiss.

And suppose your life depended on weather and soil
and lorry drivers and an army
of humble worm-like creatures called worms.

Science and Humanity. Andrew Steane, Oxford University Press (2018).
© Andrew Steane. DOI: 10.1093/oso/9780198824589.001.0001

> Now imagine you could not speak anything
> except what others can hear.
> Imagine you could not show anything
> except what others can see.
> Imagine you could not be anything
> except what others will carry.
>
> Now lightly pause, and imagine
> being carried, and carrying.
> Imagine being seen, heard, fed,
> taught, breathed, stood, awed into being.
> Look up, child, and tell me, what do you see?

A human being is not a rock, but a river. Not a statue gradually hewed or built, but a meeting point of flows. Take a look at any living body, for example: run the story in fast-forward, and you see oxygen absorbed, transported, and given off again in CO_2, and a flow of food going in, molecules taking up temporary residence before they flow on in their turn. Something similar takes place in our psychology and in everything that has happened to form us.

The only reason you can communicate is that other people can understand your language. The expression of your identity requires that others can receive and respond to it. And their identities depend on your receptiveness.

The only reason you can hear is that you have listened. To feel that you are not alone you must pay attention to the voices that address you.

Consider a large complex multidimensional network. Somewhere in this network is a node, a meeting point, that represents you. Links or threads, lots of thin ones, extend out to the nodes representing the people who made the clothes you are wearing. Other threads, thicker ones, extend to the people who taught you when you were young. Others extend to your friends and colleagues. Thick threads, like major branches, extend to the people you grew up with, especially the adults who saw to it that you were clothed and fed and taught. Others, unavoidable, extend to all the people in difficulty that you have become aware of and whose plight bothers you, or should bother you. You have absolutely not created yourself and you absolutely depend on the existence of this network, because, at the end of the day, you *are* that meeting point; there is nothing else that you are.

> Everything without exception which is of value in me comes from somewhere other than myself, not as a gift but as a loan which must be ceaselessly renewed.
>
> —Simone Weil

Born from Above

The gene does not copy.
Lying on the floor there,
it is just another molecule coiled in the dust,
as dead as charcoal.

The embryo does not grow.
Tumbling in the flow of blood,
it bags a few nutrients and comes to a close,
as inert as a clot.

The brain does not think.
Floating in a jar of nutrients,
cut off from sight and sound, from taste and touch,
it makes no connections, it is
as thoughtless as a pickled onion.

The baby does not live.
Sucking on a tube of milk,
unlooked at and unspoken to, untouched, unhugged,
unsmiled upon, it lasts perhaps a month,
perhaps a year, and then is gone,
as sad as a little coffin.

The adult does not love.
Left to his own devices,
individual, making his own way,
he lasts perhaps a decade, perhaps a century,
and builds his smoke in the wind.

He can no more live than can a baby.
He can no more think than can a brain.
He can no more grow than can an embryo.
He can no more copy than can a gene.

We are born from below—rocks give ground, plants and animals give food, biological parents give egg and sperm. And we are born from above—a mother's body gives information (negative entropy) that makes it possible for the foetus to grow; parents give eye contact, hugs, and complex information-bearing sounds called speech; all this builds up information structures in our brains. Fellow children bounce off us and let us know what is welcome and what is objectionable in our attempts at life. Teachers help us to reach maturity. Someone has talked up a storm, and the storm is us.

Genes do not copy themselves: they only get copied when they are in living bodies; they too are born from above. Brains do not make thoughts by themselves: they are only able to do this when they are in contact with the wider world: they are born from above. Trees do not grow merely by sucking up nutrients from the soil and rain; they grow by grabbing carbon from the air, and in order to do this they need a source of negative entropy: the sun. People do not live on food alone; they live on every information-bearing signal that proceeds from the network that holds them. But a network cannot generate information. The information is either fed into it continuously or was hidden there in vast quantities at the outset, or both.

In fact it is both, but the word 'information' does not capture the situation very fully; I am using it as a way in, to give some initial sense of the idea to readers unfamiliar with it.

The information or *opening* by which we live, which we absolutely require in order to learn and grow, comes partly from the resources hidden in the physical universe, and partly from a process of creativity which remains deeply mysterious from a scientific point of view. This involves a human contribution—our willingness to learn and share—and a contribution from another source, a source which in some respects does not need to do anything at all except say what things are: declare their goodness or badness; be the standard and statement of what is good and what is not good. But we find, in fact, that this influence, when experienced by ourselves, has a more active feel, providing inspiration, encouragement, and challenge. When we have an insight, into mathematics for example, or into music, then there is a sort of partnership going on, in which the truth affirms, or gives to us the opportunity to know, that the insight was indeed valid. We experience this as a profound sense of beauty. It need not feel more personal than that, but this is one aspect of the creativity that goes on. In contrast, the aspects of *encouragement* and *challenge* are experienced as somewhat more personal, as if we are being addressed.

To this we can add the experience of forgiveness. It is only available to anyone who has expressed the sense of being in need of it, of course, but, to the extent that we can manage to express such a sense honestly, then the forgiveness happens and the relationship goes forward. It is a relationship which cannot be pinned down and analysed because now we are doing business with our *other* parent, the permanent holder of value which is not about what we can dissect and pass judgement on, but is all about what we can aspire to and trust.

When people say they do not, in all honesty, believe in God, it is because they do not realize that they are using the word 'God' in the wrong way. In the modern world the word is so misused that it amounts to a redefinition, often into something paltry or superfluous. The result is that when the word is used lightly, the meaninglessness is reinforced, and when people speak more carefully, they are liable to be misunderstood. We need to find other ways of speaking, but it is difficult to do.

First there is a quality that is like a form of information, which makes the world what sort of world it is; then there is another quality that shapes what the world may become. The first is lying in vast quantities at the root of the physical world, a primordial *word* defining what the universe is in its basic nature. The second is like a continual act of creativity.

The basic nature of the world includes the potential not just to change but to develop, to support new structure which is not just different but capable of expressing richer meaning. Its temporal and incompletely controlled nature has allowed the universe to *become*; that is, to develop structures which were allowed by the action of such laws as gravity, quantum electrodynamics, and so on, but which were not dictated by them. The universe has freely explored the possibilities of what can be, and has realized many of them. This is what I presented in the early chapters of this book. But in that realization, what has actually happened? What is the truth of it? Is it all a massive farce? Is it just epic tragedy with moments of black humour? Or is it the growth of something marvellous and precious?

We find ourselves in a position of being able to ask such questions, and although we don't need full answers, we do need some sense of in what direction the answer may lie, in order to prevent ourselves living by egoism (except for the people who want to do that, I suppose, but it is the road to isolation and death). Thus, in contrast to other animals, we find ourselves somewhat in the position of assessors and steerers of the processes on planet Earth (and of the processes within our slight grasp beyond the Earth). We deeply need a sound sense of what is going on, because otherwise we will destroy ourselves. This strong statement is not an exaggeration. Truth is something which has to be worked at, and if it is neglected then things break apart. People begin to lose hope, or lose a sense of their own worth, or else they fall in the other direction and assert themselves over their contemporaries or over their descendants. Grotesque injustices pass unobjected to, mental health deteriorates, communities fall prey to prejudice, species are driven to extinction.

Is it fundamentally the case that it is best for each individual to acquire what money, sexual partners, and public accolades they can? Or is the truth of the world something radically different from this? Something in which meek people and peacemakers discover the only riches worth having?

As people have formed and handed on better answers to such questions, through their poetry, music, revered writings, and lives, another component of the becoming of the universe has been on show. This is the 'other' component in the sense that it is a process which is perfectly genuine but which cannot be adequately grappled with by science alone. The process is suitably called *inspiration*.

One must speak carefully here. This is neither like a science fiction scenario nor like a folk religion scenario. Science fiction imagines possibilities such as the whole universe being a sort of computer simulation within a larger universe; folk religion imagines the same type of scenario but expressed in old-fashioned language or in a more romantic style: a super-being living in a palace called 'heaven' and sending messages. What thoughtful and sensitive religious expression is doing is saying that this present world *is itself* the vehicle of expression of what matters. The refrain of the Bible is not 'I will pull strings for you' but 'I will be with you'.[*] Physical reality does not have, and does not need, any science fiction programmers, nor any super-beings who could sidestep the consequences of what goes on in the world. What it does need, and has, is a support of a different kind, one that creates the bridge from maths to electrons, and from love to persons.

The relationship between the inspirer and the physical world is one which respects the integrity of the latter as something which should not be coerced. However, this does not imply that there can be no living relationship between them. We experience and enact our own freedom to shape the world, and this is possible because there is a certain room for freedom within the world. It is perfectly coherent to assert that that

---

[*] The first attitude does occur in places, but as things develop it largely—almost entirely—gives way to the second.

same freedom is enjoyed by God. But we must get used to the fact that if this action respects the natural world, then it does exactly that, and therefore it will almost always be possible for us to say 'it just happened'. To recognize God is not to recognize mere power or sophistication; it is to respond to meanings that are fundamentally about love, and such a response is always about trust, not cagey defensiveness.

Recall the phrase from R. S. Thomas that I quoted in Chapter 14: it is 'the plain facts and natural happenings' that we interpret when we recognize God's contribution. An important example from recent decades is, I would say, the end of apartheid in South Africa. This was brought about by a human effort that amounted to a decision by ordinary people to embody and enact what love and justice requires. The ability to recognize what love and justice amount to was given them by God through slow but steady historical processes and occasional great steps forward when people are more receptive to inspiration, or simply to the constant challenge which perfect goodness always presents. The will to respond to such a challenge rests, ultimately, on a sense that we and others are worth something. This is a precious glimmer of an intuition that we are both lovable and loved, while also being proper subjects of an assessment which must find many of our attitudes to be objectionable.*

I used the word 'information' above in order to help the reader get some initial sense of what I am writing about, but I don't intend the word to have only its impersonal connotations. In the end what we need, and what we get, is love. As Joseph Ratzinger (Pope Benedict XVI) has put it, 'Man lives on truth and on being loved: on being loved by truth.'[54] This is not an expression of hope that something will happen, it is a statement of the fact of what is happening.

To understand this, a very helpful guide is the Jewish philosopher Martin Buber.

Here is a quotation that has been attributed to Buber:† 'When two people relate to each other authentically and humanly, God is the electricity that surges between them.' This statement will get mixed

---

* One should not neglect to mention that certain smug versions of Christian thinking were implicated in the origins of apartheid. A committed and more profoundly Christian effort was pro-active in overcoming it. However, my point in this paragraph is independent of any particular religious affiliation.

† I have not been able to locate the source, but it is certainly the kind of thing that Buber would say.

reactions. Some will feel a surge of recognition, saying, 'yes! I know what you mean!' (that is how I myself feel about Buber's statement). For others it will evoke suspicion: either the suspicion that a religious word has been brought in to give false colour to a human situation that is already precious without that colouring, or else the suspicion that an ordinary human situation has been used to demote and abuse a precious religious word. In either case, the way to respond is, I think, to make more effort to try to discover what Buber himself meant.

Here is what I think Buber is doing with a statement like this.

He is making an effort to help people get a sense of what some religious language is about. Make no mistake, Buber is under no illusion about what passes between people when they meet. He knows that no calorimeter will register anything literally passing between them like an invisible laser beam. He is also not enamoured of the use of the word 'energy' that you find in quack medicine and New Age mysticism. But he is thrilled by the experience of authentic encounter between person and person, and also convinced of its profound role in what our very existence is. He has argued carefully and at length about this in his professional work. So he is making reference to that body of work. And he is not using the word 'God' out of a wish to be perverse, or as an atheist trying to move beyond religion or reduce theism to deism. Rather, he is performing the poetry which authentic religious speech is and always must be.

This is just a single phrase, a single sentence, so one should not over-interpret it, but I want to comment on the use of the word 'is' here. Buber does not say God is *in* the dynamic that fuses a link between people; he says that very dynamic *is* what we fundamentally live in or by. The metaphor of 'electricity' also should not be misread; it suggests an impersonal nature to God, but in fact Buber's considered view was that in the overall experience of the breadth of our existence, what we encounter is, taking all into account, an all-embracing 'Thou' not 'It'. For example, he writes in *I and Thou* (1923):

> Some would deny any legitimate use of the word God because it has been misused so much. Certainly it is the most burdened of all human words. Precisely for that reason it is the most imperishable and unavoidable. And how much weight has all erroneous talk about God's nature and works (although there never has been nor can be any such talk that is not erroneous) compared with the one truth that all men who have addressed God really meant him? For whoever pronounces the word God

and really means Thou, addresses, no matter what his delusion, the true Thou of his life that cannot be restricted by any other and to whom he stands in a relationship that includes all others.

So the shorter statement evoking the surge of 'electricity' is not a statement of atheistic nor deistic theology, it is simply a way to help people go forward a little. It helps by conveying, in indirect language, the sense of 'oomph', of sheer *is-ness*, that we sometimes experience, and letting us know that insofar as this is found in an experience that is both authentic and personal-encounter-related, then the foundational and personally encountered reality that we call 'God' is forging one of the links that make us.

There remains the fact that we need a little more in order to get past atheism without resorting to guesswork: we need God to announce and show God's own nature through some act that is intelligible to us, that speaks our language, and such that an honest person can recognize it. This has happened on our planet in proportion to people's ability to hear and understand.

We are all of us all the time saying, by our choices, what we think matters, and equally we are continually receiving what truth we can. When *that which is enduring and most valuable* is expressed more fully in human affairs, it requires our willing partnership. That is to say, God's announcing of God's own nature is not a separate act from our hearing; these are two aspects of a single relationship. This is why, I suggest, it has happened so painfully slowly. But it has happened, through various self-giving and thoughtful people and the communal response to them. It seems to be largely the case that in order to prove its own validity, truth has to speak from a position of weakness, not power. Its power to convince must lie in its truthfulness alone. This suggests that the most important things we need to hear will not be shouted at us. Truth will come in humble clothes.

I will now say a little about the events of first-century Palestine; events that in the first instance barely registered as a blip on the radar of imperial Rome, and yet which directly led to a major transformation not just of human society but of the very way we conceive of what a person is. Because that transformation has worked its way deeply into human culture, modern people may not realize how much we have all benefited from it, and how different our lives would be if it had not happened.

Palestine has an interesting and arguably unique geographical and historical location, because it was for a long time a crossroads between energetic imperial powers vying for supremacy. It was repeatedly overrun, and yet was home to a people and culture which retained its sense of identity and outlasted every civilization that tried to crush it. What eventually took place there might look like a small thing, certainly a humble thing, but it had the transformative power of a modulation or a lens or a seed. A modulation or key change in music, in the hands of a great composer, serves to transform the effect of the whole piece; a lens allows one to see the same things in a new way, by bringing together the parts differently; a seed encodes the possibility of something much larger than itself.

Imagine, if you will, a man raised by a jobbing builder in a village community in a small country much fought over and currently under military occupation. A country not unlike some in the Middle East to this day. Building on the remarkable history of his people, this man elects to do something powerful in order to oppose the widespread injustice of the systems of the world, and offer a positive alternative. He adopts as his medium of expression not the written word, nor international politics or armed revolt, but sharing his life with whatever ordinary people he happens to be alongside, and from that base beginning to act more and more publicly, either encouraging or challenging what he finds. This is, roughly, the way the Christian movement got started, and it did bring about a key change. The key change brought about by Jesus of Nazareth asserts that a supposedly worthless victim of casual brutality might actually be a better representative of human and divine glory than the great athletes, generals, intellectuals, and emperors of the world. It is hard for us now to grasp quite how completely the movement which sprang up from this brought in a liberation in the very way people thought about themselves and each other. The emperor was no longer divine; the master was no longer above the slave; mercy was free not earned; temples were optional; one's worth and place was not based on human descent but on a direct worth as a child of God in one's own right. We are all beginning to see this now; it is the theme of many of our favourite films, novels, and plays. I am not saying that such insights were all completely absent elsewhere in human history, not at all; I am saying that here they were brought into sharp focus and acted upon—sometimes at great personal cost. By the lens offered in Jesus's sermons and parables and the Jewish history that he drew on, we come to see that we don't need to worry about power, we just need to pay attention to what is really going on in the world, right under the noses of, and unnoticed by, the rich and assertive, when one humble person responds generously to another (science, at its best, is a part of this).

Finally, there was the event publicly attested to within weeks and at length and at great cost by those who witnessed it: a human being given physically demonstrable, lively yet surprising and hard-to-describe new life after he had died, 'a new form of physicality for which there was no previous example and of which there remains no subsequent example.'[84] Of course this is an extraordinary and contentious claim, but it will earn attention in proportion to the character of the man it is claimed about, and the largeness of vision associated with it. This is not about mere continuation, but about something bolder and breathtaking. As Ratzinger puts it, 'he was transformed through the Cross into a new manner of bodiliness and of being-human pervaded by God's own being.'

I will not elaborate on this further here,* but it is important to include it because it represents *sufficient* evidence that reality extends beyond the ordinary processes of the natural world. Note also that it is not about political and military power; it is a revolution carried out by quietly spoken words and the glimmer of merriment in love's eye. It also indicates, of course, that the person singled out in this way is one to whom we should pay a great deal of attention. According to what we learn from that person, this is all the sign we need or will get. The response is up to us.

It does not take more than a moment's thought to see that the event I just alluded to was a one-off. Even if it happened as the witnesses report, it does not change the fact that in almost the whole of human experience, there are no miracles. Human history is not a sequence of miracles, but a muddled and vibrant complex of passions and pain, and it is often a story of seeds that begin movements, and the movements that result. In all this God is not absent nor a spectator but a player who has chosen not to overwhelm, and who has often chosen the weak and supposedly foolish things of the world in order to bring about change.

---

* Hume's point about reliability of human testimony is not sufficient to make the reports incredible, because if the whole context of our lives is in fact configured a certain way, a way that is itself coherent, then the reports are perfectly credible. To make the case for this in full requires a lengthy argument, however. This footnote merely signals that such an argument can be made in a reasonable way.[83]

For people with a keen sense of scientific integrity, it can be very helpful if we can give some account of how God's freedom to act afresh might be accommodated without distortion in the physical world. I would say that the main point is that if humans have some freedom, then so does God, and if humans do not have genuine freedom, then theism is wrong. However, I will offer a further thought which I have found helpful. It is an analogy with the way orchestral music is performed in the classical tradition. The music is scored, and yet the quality of the whole depends on a large number of improvised choices made by the individual musicians, as they adjust the volume, tone, timbre and duration of each note. In a similar way, a large number of small adjustments in physical processes can contribute to the development of the world in significant ways without overturning the patterns we call laws of nature. This is not manipulating the music of the world, but joining with and shaping it in a sympathetic and creative way. Nor should it be thought of like a laser scalpel coming in against a person's will; it is a shaping influence at the boundary of our personhood, one which meets with us as people, not collections of neurons, and which we can resist or not as we choose, because a complete power of veto on both sides is inbuilt into the very nature of personal interactions.

In this chapter I have described the process of our growth as if we were each passive in it, but of course we are not. We have, by virtue of mechanisms or laws that remain far from understood, something we call a *will*, a decision-making capacity. The exercise of this will contributes to our own development and to the development of others. This decision-making capacity is close to the heart of who we are, and this too is born from below, and from above. That is, we have to *embrace* the truth of our situation, by an exercise of the will, in both directions. We are children of the natural world—supported and embodied in a set of physical processes—and we must decide to recognize and accept this. We are equally the children of another parent—supported and informed by a larger reality that calls us—calls us by name, knows us— and we must *decide* to recognize and accept this, because until we do we are failing in our duty to truth, and perpetuating a contradiction deep in ourselves.

That contradiction will not always be apparent, but it persists in a sort of *thinness* of our existence, a certain evasiveness or lack of solidity. It emerges in a mismatch between our best vision of how to behave, and our actual behaviour. That gap reveals that the vision was not truly owned or embraced by us after all; we turn out to be devoted to something else, some sort of compromise. We feel that equal opportunity is a good thing, but we hesitate when it implies a drop in security for ourselves. We are often more eager to pay for an upgrade to a mobile phone than to pay for clean drinking water in a distant village. Thus our priorities are not our own. Our very will is in conflict with itself.

Occasionally we may experience glimmers of the fact that we are not just slightly off-base but profoundly desensitized to the enormous injustice which goes on daily in the world, and whose causes we are caught up in. Or else we may live in danger of despair and of losing our sense of the profound beauties of the world. To become more awake is to become more aware of the fearful possibility that we are profoundly attached to things that do not truly matter, such as acclaim, and unattached to things that do, such as universal justice. We absolutely require a way of tethering our life's journey onto something that will save us from this vacuous alignment and growing unreality.

For great amounts of time this nagging sense of unreality is suppressed by either worry or enjoyment. One can keep oneself entertained by music, sport, parenthood, sex, and by the wonderful adventure of scientific discovery, but none of these address our innermost being and the contradiction that exists there. The hunger for sheer reality, for a centred wisdom and loving-kindness that amounts to complete '*human being-ness*', if I can put it like that, is in our heart, and it is a precious hunger because it can be satisfied. But it is the hunger for a peace which the world cannot give. It is the precious sense that our existence is a gift, and a reciprocal giving. Ultimately we gain our life by being willing to give it up. In the end the trouble with impersonal definitions of ultimate being is that they leave us too much in charge of our own identity. They say that we can have opinions about truth, but truth cannot have opinions about us. This is not the way to realize humanity in fullness, however; it is an attempt at self-definition which has to be abandoned in favour of a relationship of mutual trust, however tentative on our side—a willingness to be a child of our other parent; our truest parent.

Lilies

Maybe this is my point:
whatever God is, He is not that:
Not that hunkered-down being you can point to, over there,
and find that it is not there;
Not that invisible printer on the waves,
walking in a mysterious way,
calling black white to poor swimmers in despair.
Nor yet what anyone ever saw through hate's narrow eyes,
nor a sneaky name for human hope.
Not the property of those who think they carry their reality around
  with them.
Not the rubbisher of unbelief, nor its comforter.
Not an adornment, not an accessory.
Not the Ideal that you can set out
on philosophy's white sheet.

None of this, but, perhaps,
One who speaks frankly in the root's eloquence;
One who stirs whale-song in the heart's ocean;
One who salmon-leaps from hand to helping hand
and stands weeping the sun's irrepressible rays.

Not anything dismissable but
One who really could clothe a human soul
as a lily is clothed.

22

Witnessed to

Try this as a spiritual exercise.

Imagine as best you can how you might feel towards your child, if you were a parent with a child aged around ten or so. That is, a child well launched on the way to adulthood, but not yet there. You may be a parent in that very position, or have been in the past. If not, then it may help to think of a specific child very well known to and loved by you. Just ponder for a few minutes what thinking about such a child is like—how they enter into your thoughts.

Now, I think, if you have a reasonably sound and generous sense of these things, then you will not think of your children first and foremost either in terms of their capabilities, nor in terms of their faults. No, when I reflect on my children, and I think when others reflect on theirs, we find we are first and foremost interested in them just for who they are. We love them just as people with characters. We do care about how they behave, but that is not our first care. We have a deep sense of connection to, and bear a little bit thrilled witness to, simply a person with their own unique and precious character. The way those characters emerge in particular actions or capacities is secondary. If they are good at kicking a football or solving a mathematical puzzle or playing the piano, then all well and good, but that is not the first thing that springs to mind. If they are poor at tidying their room or being courteous to their sister or getting on with their homework, that is something that needs attention, but it is not the first thing that springs to mind. No, what springs to mind first of all is just the child in himself, in herself. The one we are coming to know.

That is the end of the spiritual exercise. Here comes the application. Following some other teachers, but primarily taking the lead offered by Jesus of Nazareth, Christians speak of God as a 'Father'. That is, we say that the non-contingent support of reality, the thing which does

Science and Humanity. Andrew Steane, Oxford University Press (2018).
© Andrew Steane. DOI: 10.1093/oso/9780198824589.001.0001

not depend on anything else, because it is not a thing but that which gives rise to things—we say that this deepest and most creative reality is like a parent, or can be helpfully and truthfully (though incompletely) thought of by us as like a parent.

So, when we say this, what are we saying? We are saying that the relation between ourselves (that is, all carriers of personhood, whether on Earth or elsewhere) and the unconditional source of existence is like the relation I invited you to think about just now. You shouldn't buy the claim unless there are sufficient indicators that it is trustworthy, of course, but my aim here is simply to clarify what the claim is. It is the claim that the root of being—that which defines what existence is, and is encountered by the human body and brain in the here and now— is not like a stroke of luck, or like a clutch of mathematical principles or ethical principles, nor like a monarch or a potentate, nor a usable force, nor completely unknown and unknowable. That which defines what existence is, and is encountered by us directly in and through our personal identity, is like one who witnesses to and affirms who we are. We are recognized not primarily as good and capable, nor as wicked and incapable, but as simply ourselves.

That is the reality we find ourselves in.

Three Hundred Years of Ridicule

Many eminent doctors stood about the body,
proclaiming it dead. They proclaimed it dead.

But if it is dead there is no need to
keep on expostulating, because it is dead.

Those fools singing in the cathedral have no eyes of sense.
Don't they know when something is dead?

But at least spare a thought for the mourners,
grief-paralysed for one whom they hoped was not dead.

Some crept under the barrage of ridicule to
tend to the corpse, but it had gone.

Gone.

They found themselves in a strange garden
which did not feel at all dead.

And no thing yet someone said 'take heart:
if you live, there I will be not dead but living,
living,
living'.

APPENDIX
Boyle's Law

In this appendix I give the argument about Boyle's law for a gas, which was used as an example in Section 3.4. I first present the gist of it for the non-expert reader, then I give details for the expert reader.

The goal is to discover how much it is possible to deduce about the behaviour of a gas from general thermodynamic principles, without bringing in a detailed model of the parts of the gas. That is to say, we accept that a gas is made out of many parts (such as individual molecules), but we would like to explore what is the least we can say about the parts, and still get correct statements about the whole. This aim is quite extreme: we want to suppose almost complete ignorance of what these 'molecules' or whatever are like. We don't necessarily assume them to be like little bitty things that whizz around between the walls of the container of gas; for all we know each part might extend like a transparent spring or a vibration throughout the whole container of gas, or maybe it is something else entirely. We will need to suppose a little, however, such as that these parts, whatever they are, cannot appear and disappear. The specific assumptions made in the argument will be spelled out in what follows.

The argument proceeds in two steps. First we bring in general thermodynamic reasoning, which tells how pressure and volume are related to another property called the Helmholtz function. This first step says nothing to distinguish a gas from other things such as a liquid or a solid; it simply asserts what kind of physical quantity pressure is, and how it relates to energy and entropy. The relation to energy and entropy is written down in terms of the Helmholtz function merely because this is a convenient way to do the mathematics.

The second step in the argument is to provide further information about the Helmholtz function, so as to find out how it behaves for the specific physical system we have in mind, namely a gas. This logically requires that we bring in some sort of further information, enough to distinguish the gas from other things such as a liquid or solid. Our question is, how much further information is required?

It turns out that we can manage with just the following:

1. The gas is composed of a fixed, large, number of parts.
2. These parts interact with one another only weakly or intermittently.
3. The parts all are the same as one another, in a strict sense: they are mutually indistinguishable.
4. The parts are not entirely unlike other simple things, in that they have states of motion that could in principle be counted (but we don't need

to know what those states are, or how their energy relates to their momentum, or things like that).

The 'parts' referred to here could, for example, be individual molecules, but, as we already noted, the argument does not require that. The argument requires almost no knowledge of what these 'parts' are. They could, for example, be not particles at all, but some sort of vibration or wave, as long as there is a sense in which they can be counted, so that the total number of them can be fixed. We do not require any statement about the physical nature or properties of the parts, nor how one part affects another, nor how they change with time, except the general constraints that are expressed in the laws of thermodynamics. That is, the net result of the motions of the parts, whatever they may be, is that energy is conserved, and in equilibrium the overall state has the highest entropy that is available under the restriction that the whole system is isolated. (For experts: the salient fact is that the derivation does not require the functional form of the single-particle partition function z_1. It suffices to know that z_1 is independent of the number of particles at given temperature and volume; the details are given below.)

The rest of the argument consists of mathematical steps which invoke some general statistical methods. The statistical methods (called statistical mechanics) are also based on the idea that large things are composed of lots of small parts, but they do not require detailed information about the parts, except that mentioned in the fourth item above.

Using only the above assumptions and methods, Boyle's law can be derived (to be precise, one derives both that pressure is proportional to temperature and that it is inversely proportional to volume at any given temperature).

To repeat, we thus learn about a gas without needing to first find out about the bits and pieces that make a gas, because the large-scale, symmetry-like ideas (energy, entropy, and the general concept of a state space) are sufficient to tell us how it has to be, if thermodynamics holds, and if a gas is a loose collection of many indistinguishable, countable parts.

A small further ingredient suffices to produce a model which captures, qualitatively, the liquid behaviour also, and the main facts about the transition from liquid to gas. This further ingredient is the idea that the parts of the gas take up a little volume of their own, and slightly pull together when they are separated far apart.

Detailed Derivation

I now provide the mathematical details. This section can be omitted by readers content to trust that I have not got this wrong. It has been checked by anonymous peer review by fellow scientists with the relevant expertise; I include it here in order to allow a wider assessment. The mathematical correctness is not really in any doubt; what can be debated is the nature of the assumptions and the significance of the conclusion.

We begin by quoting the fundamental thermodynamic relation for a simple compressible thermodynamic system composed of a single chemical species:

$$dU = TdS - pdV + \mu dN,$$

where all the symbols have their usual meanings, and in particular μ is the chemical potential per particle and N is the number of particles. If one prefers not to invoke the concept of particles, then one may instead use moles and the molar chemical potential; the argument is not changed. Note, the use of the word 'particle' here does not involve any assumption of their nature beyond the notion of a conservation law for N. By considering that U, S, V, and N are extensive, and T, p, μ are intensive, one obtains the Euler relation

$$U = TS - pV + \mu N.$$

Now introduce the Helmholtz function $F = U - TS$. We have $dF = -SdT - pdV + \mu dN$, and therefore

$$\mu = \left.\frac{\partial F}{\partial N}\right|_{T,V}.$$

Using this expression for μ, and writing the Euler relation in the form $F = \mu N - pV$, one obtains

$$pV = N \left.\frac{\partial F}{\partial N}\right|_{T,V} - F. \tag{A.1}$$

This is a general thermodynamic relationship which applies to any simple compressible system having a single type of matter, in thermal equilibrium.

Next we introduce the relationship between Helmholtz function and partition function which is a standard result in statistical mechanics. In the canonical ensemble one has

$$F = -k_B T \ln Z_N \tag{A.2}$$

where Z_N is the partition function of the system (here, the gas). This is a general relationship which follows from the principle that the equilibrium state is the one that maximizes the entropy. It is not dependent on any particular model of the structure of the system in question, except the general notion that there is a state space (whether quantum or classical or other) that is sampled evenly by the system.

Now we introduce an assumption that makes the treatment somewhat more specific. We make the claim that for an ideal gas in the low density limit,

$$Z_N = \frac{z_1^N}{N!} \tag{A.3}$$

where N is a large number and z_1 does not depend on N at given T and V. This mathematical assertion is motivated by two assertions about the physical situation: we assume that the system we are dealing with is made of a large number N of weakly interacting parts, and those parts are strictly indistinguishable. The phrase 'weakly interacting' here means that the states available to one part are not affected by the other parts, and nor are transition probabilities among them, except through the conservation of energy, but there is a slight interaction that suffices to allow the system as a whole to explore its energy surface and come to equilibrium. The term 'strictly indistinguishable' means that when the parts are rearranged among themselves, we do not have a new physical state. Under these assumptions (A.3) follows, and z_1 is called the 'single-particle partition function'. However, note that we say 'each part' here, not 'each particle' or 'each wave', because we make no assumption about the nature of these parts, beyond their weak interaction and their indistinguishability. In the following we will not need to know anything about z_1, except its independence of N at given T and V.

Equations (A.2) and (A.3) imply that the Helmholtz function is given by

$$F = -k_B T \ln(z_1^N/N!)$$
$$= -Nk_B T (\ln z_1 - \ln N + 1) \tag{A.4}$$

where the second line is accurate in the limit of large N. Therefore

$$\left.\frac{\partial F}{\partial N}\right|_{T,V} = -k_B T(\ln z_1 - \ln N). \tag{A.5}$$

Using equations (A.4) and (A.5) in (A.1) gives

$$pV = Nk_B T$$

which is the equation of state of an ideal gas.

A method of calculation much like the above, but invoking the grand potential and the grand partition function, may be found in statistical mechanics texts.[32, 39]

Note that the above calculation does not require any study of the density of states, nor the equation relating energy to momentum for the parts. This is a striking fact which merits emphasis; in order to obtain the equation of state, neither the dynamics, nor even much kinematics is required: one needs to know only very little about even what *sort* of thing a gas may be composed of.

The equation of state does not, on its own, yield the complete thermodynamic behaviour of a gas. Other features, such as the fluctuations and the heat capacity, will differ from one ideal gas to another. Nonetheless, the example illustrates the way thermodynamic reasoning functions in its own terms, drawing correct and general connections between physical properties, without the need for anything but a small number of assertions about the microscopic structure.

By invoking the partition function in equations (A.2) and (A.3), the argument does bring in the concept of a state space, and is conceptually demanding. In (A.3) it also invoked a deeply non-classical concept, namely strict indistinguishability of matter entities. This is like the indistinguishability of lumps of energy or momentum, but it is a non-trivial statement to assert that sort of indistinguishability for lumps of matter.

If one wishes to go further, and obtain a qualitative model for the vapour–liquid phase transition, then the van der Waals treatment can be employed, which is an example of mean field theory. This also requires few assumptions.

Notes

The short story or parable called 'Light' uses some ideas from particle physics. The term 'light' is questionable only insofar as it fails to make the distinction between bosonic and fermionic fields. These are, roughly speaking, what in ordinary language may be called energy and matter, but in the physics of quantum fields this distinction is blurred. The pair of counter-propagating flashes mentioned near the start are the parts of a Dirac spinor in the chiral representation. They each have a phase velocity equal to the speed of light, and their combination gives an electron moving at some slower speed. The other 'bump' is a proton; the helix is a hydrogen atom; the 'factory' mentioned later on is a living cell. And so on.

Shamayim is the transliteration of a Hebrew word that appears often in the Bible, and is usually translated 'heaven', but that modern English word singularly fails to capture the Hebrew idea.

The poem entitled 'Performance' accurately describes an exchange that took place after a lecture by Prof. Richard Dawkins, which I saw on a video clip.

The titles 'Sky Hooks' and 'Universal Acid' make reference to ideas presented by Daniel Dennett in his book *Darwin's Dangerous Idea*.

The title 'Born from Above' refers to the discourse in the third chapter of the gospel according to John. It is not my intention to state a definite preference for any particular translation, nor to suggest that the poem adequately captures the meaning of the phrase in the context of Jesus's conversation with Nicodemus. My intention is rather to show a related idea that may point one in a good general direction.

Bibliography

1. In his *Metaphysics* VIII.6, 1045a, Aristotle uses a heap as an example of a thing that is not much more than the sum of its parts, and this is right. Nonetheless, even a humble heap has properties that are about the whole not the parts. A heap of apples exhibits mathematical facts about close-packing arrangements of nearly spherical objects, for example.
2. 'The argument from design fails, then, because it is an argument from ignorance, because it confuses the final and efficient modes of explanation, and because even if it succeeded it would not prove the existence of God but of some Masonic impostor. But like other bad arguments, its defeat and death has left it to wander the world like a ghost, oppressing the spirits of those who are looking for other and better arguments.' (F. J. Martin, *Thomas Aquinas: God and Explanations*, pp. 181–2).
3. The clergy letter project: http://www.theclergyletterproject.org/.
4. P. W. Anderson. 'More is Different.' *Science*, 177: 393–6, 1972.
5. Maya Angelou. 'Human Family.' In *I Shall Not Be Moved*. Random House, New York, 1990.
6. Anon. *The Cloud of Unknowing*. John Watkins, London, 1922. Evelyn Underhill, ed.
7. Anselm. 'Proslogion.' In *Complete Philosophical and Theological Treatises of Anselm of Canterbury*. The Arthur J. Banning Press, Minneapolis, 2000.
8. St Thomas Aquinas. *Summa Theologiae*. 1274.
9. St Thomas Aquinas. *Summa Theologiae* 1.3.5 sed contra. 1274.
10. Karen Armstrong. *Fields of Blood: Religion and the History of Violence*. Vintage, London, 2014.
11. Robert Axelrod. *The Evolution of Cooperation*. Basic Books, New York, 1984.
12. E. C. Barnes. 'Ockham's Razor and the Anti-superfluity Principle.' *Erkenntnis*, 53: 353–74, 2000.
13. Wilhelm Baum and Dietmar W. Winkler. *The Church of the East*. Routledge Curzon, London, 2003.
14. D. Bonhoeffer. *Letters and Papers from Prison (1943–45)*. Fortress Press, Minneapolis, 2010.
15. J. Butterfield. 'Emergence, Reduction and Supervenience: A Varied Landscape.' *Foundations of Physics*, 41: 920–59, 2011.
16. J. Butterfield. 'Less is Different: Emergence and Reduction Reconciled.' *Foundations of Physics*, 41: 1065–135, 2011.
17. L. Campbell. *The Life of James Clerk Maxwell: With a Selection from His Correspondence and Occasional Writings and a Sketch of His Contributions to Science*. Macmillan, London, 1882.

18. Lewis Carroll. 'What the Tortoise said to Achilles.' *Mind*, 4: 278–80, 1895.
19. Sean Carroll. *The Big Picture—On the Origins of Life, Meaning and the Universe Itself*. Oneworld, London, 2017.
20. John Cornwell. *Darwin's Angel: An Angelic Riposte to* The God Delusion. Profile Books, London, 2007.
21. Shamik Dasgupta. 'Symmetry as an Epistemic Notion (Twice Over).' *British Journal for the Philosophy of Science*, 67: 837–878, 2016.
22. Richard Dawkins. *River Out of Eden*. Basic Books, New York, 1996.
23. Richard Dawkins. *Climbing Mount Improbable*. W. W. Norton, New York, 1999.
24. Richard Dawkins. *The Selfish Gene*. Oxford University Press, Oxford, 40th anniversary edition, 2016.
25. D. C. Dennett. *Darwin's Dangerous Idea: Evolution and the Meaning of Life*. Penguin, London, 1997.
26. F. Dizadji-Bahmani, R. Frigg, and S. Hartmann. 'Who's Afraid of Nagelian Reduction?' *Erkenntnis*, 73: 393–412, 2010.
27. James E. Dolezal. *God without Parts: Divine Simplicity and the Metaphysics of God's Absoluteness*. Pickwick Publications, Eugene, OR, 2011.
28. Albert Einstein. 'Zur Elektrodynamik bewegter Körper.' *Annalen der Physik*, 17: 891–921, 1905. See also A. Einstein, *The Principle of Relativity*. Methuen, London, 1923.
29. Edward Feser. Classical theism. 2010. http://edwardfeser.blogspot.com/2010/09/classical-theism.html.
30. Diego Gambetta and Steffen Hertog. 'Engineers of Jihad.' *Sociology Working Papers, University of Oxford*, 2007–10, 2007.
31. S. J. Gould. 'Male Nipples and Clitoral Ripples.' *Columbia: A Journal of Literature and Art*, 20: 80–96, 1993. Reproduced in *Bully for Brontosaurus: Reflections in Natural History*. New York: W. W. Norton.
32. Tony Guénault. *Statistical Physics* (second edition). Kluwer Academic, Dordrecht, 1995.
33. James Hannam. *God's Philosophers: How the Medieval World Laid the Foundations of Modern Science*. Icon Books Ltd, London, 2009.
34. John Heilbron. *Galileo*. Oxford University Press, Oxford, 2010.
35. David Hume. *Dialogues Concerning Natural Religion*. 1779. Available at http://www.gutenberg.org/files/4583/4583-h/4583-h.htm.
36. J. Vernon Jensen. 'Return to the Wilberforce–Huxley debate.' *The British Journal for the History of Science*, 21: 161–79, 1988.
37. Stuart A. Kauffman. *Reinventing the Sacred*. Basic Books, New York, 2008.
38. B. Leftow and Davies B., eds. *Aquinas: Summa Theologiae, Questions on God*. Cambridge University Press, Cambridge, 2006.
39. F. Mandl. *Statistical Physics*. Wiley, London, 1971.
40. F. J. Martin. *Thomas Aquinas: God and Explanations*. Edinburgh University Press, Edinburgh, 1997.

41. Alastair McIntosh. *Soil and Soul*. Aurum Press, London, 2001.
42. Tom McLeish. *Faith and Wisdom in Science*. Oxford University Press, Oxford, 2016.
43. Tom C. B. McLeish, Richard G. Bower, Brian K. Tanner, Hannah E. Smithson, Cecilia Panti, Neil Lewis, and Giles E. M. Gasper. 'History: A Medieval Multiverse.' *Nature*, 507: 161, 2014. http://ordered-universe.com.
44. Kei Miller. *The Cartographer Tries to Map a Way to Zion*. Carcanet Press, Manchester, 2014. Lines from poem xxi (untitled).
45. Simon Conway Morris. *Life's Solution: Inevitable Humans in a Lonely Universe*. Cambridge University Press, Cambridge, 2005.
46. William Edward Morris and Charlotte R. Brown. 'David Hume.' In Edward N. Zalta, ed., *The Stanford Encyclopedia of Philosophy*. Spring 2016 edition.
47. Ernest Nagel. *The Structure of Science: Problems in the Logic of Scientific Explanation*. Harcourt, New York, 1961.
48. Alice Oswald. 'Mountains.' In *The Thing in the Gap Stone Stile*. Faber & Faber, London, 2007.
49. Rosa Parks and Gregory J. Reed. *Quiet Strength*. Zondervan, Grand Rapids, MI, 2000.
50. D. Z. Phillips. *The Concept of Prayer*. Routledge, London, 1965.
51. A. Plantinga. 'Darwin, Mind and Meaning.' *Books and Culture*, May/June, 1996.
52. Alvin Plantinga. *Where the Conflict Really Lies*. Oxford University Press, Oxford, 2011.
53. Angel Rabasa and Cheryl Benard. *Eurojihad: Patterns of Islamic Radicalization and Terrorism in Europe*. Cambridge University Press, Cambridge, 2015.
54. Joseph Ratzinger. *Jesus of Nazareth*. Doubleday, New York, 2007.
55. P. Richmond. 'Richard Dawkins' Darwinian Objection to Unexplained Complexity in God.' *Science and Christian Belief*, 19: 99–116, 2007.
56. W. Rindler. *Relativity: Special, General, and Cosmological* (second edition). Oxford University Press, Oxford, 2006.
57. Byron Rogers. *The Man Who Went Into the West: the Life of R. S. Thomas*. Aurum Press, London, 2007.
58. J. Sacks. *The Great Partnership*. Hodder, London, 2012.
59. Marc Sageman. *Understanding Terror Networks*. University of Pennsylvania Press, Philadelphia, 2004.
60. P. Sanlon. *Simply God: Recovering the Classical Trinity*. Inter-Varsity Press, Nottingham, 2014.
61. W. Shakespeare. *Macbeth*. Act II, scene iii.
62. Rupert Shortt. *God is No Thing: Coherent Christianity*. C. Hurst & Co., London, 2016.
63. L. Sidentop. *Inventing the Individual: The Origins of Western Liberalism*. Allen Lane, London, 2014.
64. Dava Sobel. *Galileo's Daughter*. Fourth Estate, London, 2008.

65. Francis Spufford. *Unapologetic: Why, Despite Everything, Christianity Can Still Make Surprising Emotional Sense*. Faber & Faber, London, 2013.
66. Richard Swinburne. *Is There a God?* Oxford University Press, Oxford, 1996. pp. 95–113.
67. Richard Swinburne. 'God as the Simplest Explanation of the Universe.' *European Journal for Philosophy of Religion*, 2: 1–24, 2010.
68. R. S. Thomas. *Frequencies*. Macmillan, London, 1978.
69. R. S. Thomas. *Between Here and Now*. Macmillan, London, 1981.
70. R. S. Thomas. *Destinations*. The Celandine Press, Shipston-on-Stour, 1985.
71. R. S. Thomas. *Counterpoint*. Bloodaxe Books, Newcastle, 1990.
72. R. S. Thomas. *R. S. Thomas: Collected Poems 1945–1990*. Phoenix, London, 2000.
73. Stephen Tuck. *Beyond Atlanta: The Struggle for Racial Equality in Georgia, 1940–1980*. University of Georgia Press, Athens, GA, 2003.
74. P. Van Inwagen. 'Is God an Unnecessary Hypothesis?' In A. Dole and A. Chignell, eds, *God and the Ethics of Belief*. Cambridge University Press, Cambridge, 2005.
75. Roger Wagner and Andrew Briggs. *The Penultimate Curiosity: How Science Swims in the Slipstream of Ultimate Questions*. Oxford University Press, Oxford, 2016.
76. K. Ward. *A Vision to Pursue: Beyond the Crisis in Christianity*. SCM Press, London, 1991.
77. Simone Weil. *Simone Weil: An Anthology*. Penguin, London, 2005.
78. David Wilkinson. *Science, Religion, and the Search for Extraterrestrial Intelligence*. Oxford University Press, Oxford, 2013.
79. R. Williams. *Silence and Honey Cakes: The wisdom of the Desert*. Lion, Oxford, 2003.
80. R. Williams. *Faith in the Public Square*. Bloomsbury, London, 2005.
81. R. Williams. *Meeting God in Mark*. SPCK, London, 2014.
82. L. Wittgenstein. Lecture on Ethics. 1929.
83. N. T. Wright. *The Resurrection of the Son of God*. Augsburg Fortress, Minneapolis, 2003.
84. N. T. Wright. 'The Resurrection Was as Shocking Then as it is Now.' *Guardian Newspaper: Comment is Free*, Monday 3 August, 2009.

Index

acid, universal 241, 279
aesthetic 122, 124, 137, 217, 243, 245–6
agnosticism 212, 247
Albertus Magnus 127
Alexander, D. 100
algorithmic 56, 60
Alhazen (Ibn al-Haytham) 127
alien 200, 216, 221
Anderson, P. 28, 30
anteater 80–2
anthropic 226, 227
apartheid 261
Aquinas 6, 176–8, 184–5, 192–3, 196
arch 4, 5, 15, 67, 95
Aristotle 125–6, 128
arithmetic 19, 50, 60–1, 111, 149, 232, 250
art 7, 18, 105, 122, 124, 130, 193, 199, 206, 228, 232, 243
astronomy 127–9
atheism 101, 123, 137, 153, 160, 164, 186, 188, 202, 204, 207, 212, 225–8, 246, 250, 263
Attenborough, D. 224
attention 2, 4, 7, 25, 33, 54, 66, 68, 122, 129, 163, 208, 212, 248, 250, 253, 256, 265, 269
Augustine 196
Axelrod, R. 69

Babel fallacy 54, 57–64, 71, 91
Babylonians 57
Bach, J. S. 167, 250
Bacon, F. 128
Bacon, R. 127–8
Barrett, S. 167
belief 61, 186
Benedict XVI 261
Berkeley, G. 186
Bible 57, 98, 102, 128–9, 165–6, 230, 260, 279
binary logic 19, 20, 62, 95, 217
Bonhoeffer, D. 197, 199–203, 205, 209, 214
bonobo 110
Boyle's law 35, 66, 69, 273–4
Bradwardine, T. 127

Brahe, T. 130
brain 1, 22–3, 49, 50, 63, 66, 93, 97, 137, 242–3, 255, 257, 270
Brecht, B. 129
brotherhood 109, 145, 206
Buber, M. 261–2
Buchenwald 214
Buddhism 247
Buridan, J. 127
Burley, W. 127
Butterfield, J. 35

Canute 131
carbon 115, 117–18, 258
Carl, N. 90
Carroll, L. 147
Carroll, S. 18, 70
category error 7, 71, 81, 84, 86–7
chance 18, 46, 78
chaos 38, 45, 226–7
Charlemagne 126
chess 16, 17, 33
Chomsky hierarchy 59
Christian 101, 123, 125–9, 155, 178, 184, 193–4, 199–203, 206, 213, 229, 233, 247, 249–51, 261, 264
church 128–32, 213, 251
classical theism 194
Cloud of Unknowing 203
Collins, F. 100
compassion 95–6, 155, 193, 206, 231–2
complexity 38, 97, 183–4, 192, 210–11
computer 19–21, 31–3, 45, 51, 54, 56, 60, 62–3, 95, 260
consciousness 66, 244
convener 164, 192
Conway Morris, S. 100
Copernicus 128, 130
covariance 30, 65
creation 102, 132, 230
creationist 101
creativity 102, 252, 258–9

Dante 250
Darwin, C. 100–11, 149, 279
Dawkins, R. 14–15, 18, 84, 86, 90, 99–101, 110, 177, 178, 180–5, 191, 196, 198, 201, 213, 224–5, 232, 279
death 102, 111, 200, 212, 224, 259
deism 262
democracy 251
Dennett, D. 181, 183, 191, 229, 279
design 91, 100–1, 223–6, 229–30, 235
Designer 180–1, 191
deterministic 20, 51, 104
Dialogues 179, 183, 191, 199
Dirac field 140, 279
divine simplicity 192–5
DNA 46, 75, 83, 85, 103, 115–17, 191, 242
Dostoyevsky, F. 250
doubt 136, 163
Dumbleton, J. 127
Duns Scotus 127

eager gene 93, 106, 108–9
ecosystem 47, 83, 90, 102, 109–10, 138, 224, 248
education 22, 45, 99, 111, 122, 159, 161, 195, 251
Einstein, A. 26, 36, 59
electrodynamics 21, 259
electromagnetism 27, 29, 65
embodiment 28, 40, 42, 45, 52, 64, 65, 70, 72, 91, 110, 188, 202
emergence 41
empathy 75, 105, 137, 151, 192, 229, 231–2
encounter 53, 133, 139, 176, 207–8, 234, 262
Enlightenment 250
Enlil 57
entropy 30, 32–4, 41, 44, 47, 92, 94–5, 116, 273–5
eternal 192, 200
ethics 3, 15, 113, 182, 199, 201, 231
Euclid 125
Euler equations 37
Euripedes 95
Europe 125–9, 155
evidence 2, 14, 35, 56, 61, 102, 126, 130, 132, 154, 156, 161, 186, 188, 208, 210–11, 226, 235, 247–8, 265

evolution, biological 1, 18–19, 45, 69, 74, 97, 183, 244
experience 42, 71–2, 83, 102, 105, 122–3, 135, 137–9, 164, 167, 179–80, 186, 188, 191, 201–2, 204–5, 207–8, 210–12, 225, 230–1, 239, 243, 246–8, 253, 258, 260, 262–3, 265, 267

faith, *see also* willingness 99, 131, 134–7, 153–5, 160, 181, 191, 204, 219, 228, 253
Fischer, B. 88
FitzGerald, G. 26
fluctuation 31, 51
fluid 7, 35, 37, 39, 45, 118
forest 75, 89, 224
forgiveness 5, 22, 72, 167, 198, 203, 211, 221, 250, 258
fraternity 153, 166, 212
friendship 110, 190, 199, 210

Galileo 128–1
Gaudi, A. 250
Gautama 248
gene 84–5, 87–90, 94, 99, 106, 257
genus 192, 193
goal 35, 75–6, 81–2, 91, 163, 196, 202, 273
Gödel, K. 16, 61–2, 146, 149
Goethe, J. W. von 95, 141
goodness 7, 53, 139, 144, 153, 155–6, 160, 196, 202, 211, 223–4, 226, 229, 234, 244, 253, 258, 261
gospel 133, 199, 248, 279
Gould, S. J. 100, 108, 189
grammar 60
Grosseteste, R. 127–8

handedness 28–30
Hawking, S. 14
Hayyān, Jābir ibn 127
heaven 133, 140, 260, 279
Hebrew 57, 102, 165, 279
helium 55–6, 58, 68
Heytesbury, W. 127
Hilbert, D. 61–2
Hippocrates 125
history 59, 99, 110, 124, 155, 249, 264

Index

hope 94, 136, 158, 160, 164, 166–7, 200, 207, 211, 219–20, 227–8, 232, 234, 246–7, 260, 268
Hume, D. 5, 176, 265
humour 234
Huxley, T. 132
hypothesis 24, 35, 137, 184, 186, 189, 195–6, 198–9, 201, 208, 213

indistinguishability 276–7
injustice 90, 122, 139, 160, 167, 227, 229, 233, 264, 267
inspiration 115, 118, 244, 258, 260–1
intimacy 139, 188, 190
Isaiah 178
Ishaq, Hunayn ibn 127
Islamic Golden Age 126

Jacobs, C. 90
Jensen, J. 132
Jesus 197–9, 201–3, 206, 209, 212–13, 248–9, 251–2, 264, 269, 279
Jewish 155, 165, 249, 251, 261, 264
justice 3, 18, 71, 95–6, 139, 145, 153, 160, 164–6, 193, 196, 206, 221, 224, 227, 233, 249, 261, 267

Kepler 128, 130, 216
Khmer Rouge 159
King Lear 190–1
King, M. L. 160
kingdom of God 213, 233, 249

ladder of explanation 4, 22, 31, 34, 36, 45, 50
language 40
Laozi 95
law 34, 35, 41, 44, 45, 66, 69, 76, 126, 169, 200, 244, 273–5
Lego 59
LEP 182, 206
Lincoln, A. 202
linguistics 54, 59, 60
lioness 87, 92–4, 105
literalism 99
Lorentz contraction 26, 30, 41, 45, 65, 67
Louis, A. 100
Luther, M. 155

McCabe, H. 196
machine 3, 7, 14, 19–20, 22–3, 50–1, 59, 62, 71, 78, 82, 84, 88, 109
MacKinnon, D. 196
Magna Carta 251
Mandela, N. 88
Marriott, P. 149
Marx, K. 60
mathematics 50, 54, 60–1, 108–11, 127, 146, 194, 213, 218, 232, 234, 253, 258
Maxwell, G. 224
Maxwell, J. C. 27, 32, 45, 65, 202
mercy 22, 111–12, 145, 158, 165–6, 196, 198, 212, 221, 264
Merton, T. 203
metaphor 14, 79, 85, 106, 108, 262
metaphysics 133
Miller, K. 100, 197
mindfulness 240, 254
miracles 101, 265
misogyny 250
monkey 54–6, 58–9, 63, 65–7, 78, 84
Monty Python 91
morality 2, 18, 69–71, 90, 98, 108, 141–3, 158–61, 213, 217–18, 231–2, 243–6, 248, 250–1, 253
Mozart, W. A. 167, 211, 218, 243
Murphy, Eddie 106
music 5, 124, 137, 143, 193–4, 196, 206, 212, 228, 245, 258, 260, 264, 266–7
mysticism 203, 208, 247, 262
myth 3, 57, 128, 250

naturalism 70–1
Newton, I. 24, 36–7
Newtonian 59, 97
Nicholas of Cusa 127
Nietzsche, F. 204
nonviolence 168

objective 70–2, 122, 142–3, 156, 162, 204–5, 243, 246, 249
Occam's razor 45, 127, 137
Omnipotence 169
Oresme, N. 127
origin 57, 97, 179, 183, 203, 210–11, 213, 225

pain 101, 105, 107, 140, 160, 224–5, 229, 232–5, 248, 250, 265
Palestine 250, 263–4
Paley, W. 99–101
pantheism 248
parable 279
parent 187, 198, 211, 213, 231, 258, 267, 269, 270
parity violation 29
Parks, R. 202
Peckham, J. 127
personal 6, 95–6, 123, 136–7, 139, 179, 188–91, 196, 205–6, 208, 210, 212–13, 258, 266
personhood 6, 181, 186, 189, 202, 266, 270
Phillips, D. Z. 181
Philoponus 126, 128
Pink Floyd 167
pity 225–6
Plantinga, A. 183–4, 186
Plato 125
poem 73, 118, 140, 157, 169, 215, 222, 236, 241, 255, 257, 268, 271
pointillist 33, 76
politics 5, 68, 138, 159, 264
power 2, 57, 68, 71, 142, 161, 164, 166, 183, 204, 249–52, 261, 263, 265
prayer 234, 240, 253
program 20, 51, 62, 85, 126
proof 20, 39, 61–2, 67–8, 118, 146, 149–50
Psalm 230
Ptolemy 125, 128
public understanding of science 1, 14, 70, 90, 101
Punke, M. 234
Pygmalion 3

quantum entanglement 104
quantum field theory 13, 16, 27, 32, 35, 44–5, 94, 134, 193
quantum mechanics 32, 35, 44, 56, 59, 104, 275

randomness 18–20, 31, 46, 49, 51, 53, 91, 102–4, 117, 188, 214, 226–7, 229–30
rationality 122–3, 134, 212, 227, 247
Ratzinger, J. 261, 265

reality 1, 6, 68, 154, 164, 166, 176, 178, 187, 189, 198, 201, 226, 246, 263, 265–7, 270
reason 7, 102, 121, 134, 145, 147–8, 151, 153–6, 159, 179, 199, 205, 253
recognize 6, 43, 48, 52, 69, 87, 93, 105–6, 123–4, 144–5, 147, 153, 155, 167, 178, 182, 196, 202, 204, 251–2, 254, 261, 263, 266
reduction 40–3, 163
reductionism 40–2, 66–7, 253
Reformation 129, 252
Reid, T. 186
relativity 28, 36, 43
religion 6, 57, 99–100, 113, 122–4, 126–7, 129, 133, 158–62, 165, 177–8, 181–4, 199, 204, 211–12, 217, 245, 248, 250, 253, 260, 262
resurrection 200
Riccioli, G. B. 130
Richard of Wallingford 127
Richmond, P. 184
Rivonia 88
robot 88, 95–6, 138
root 134, 183, 259, 268, 270
rules, moral 142, 218
Russell, B. 176, 229

Sacks, J. 212
sacred 126, 252
scholarship 70, 100, 130, 158, 161, 182, 201
secular 212, 251
selfish gene 85, 106–9, 223
Seurat 33
Shakespeare, W. 95, 190
Shannon, C. 62
shoebill 49, 231
Sidentop, L. 126, 212
silence 139, 204–5, 240
simplicity 134, 182, 194, 211
slavery 52, 244, 249
social 18, 45, 48–9, 52, 56, 65–6, 71, 92–3, 95, 105, 109–10, 112, 128, 132, 161, 199, 231, 243, 249
solidarity 155, 161, 231, 235
soul 203, 268
space-time 3, 30, 41, 134, 226
spinor 193, 279
spirituality 4, 7, 122, 168, 203, 217, 239, 250, 269
Sumerian 57

Summa Theologiae 177, 185, 192
superfluidity 55–6, 59
superfluity 137, 185, 206, 208
supernatural 123, 133, 164, 189
Swineshead, R. 127
syllogism 35, 66, 69, 190, 197
symmetry 19, 23–31, 33–4, 37, 40–1, 44, 48, 52, 65–6, 68, 72, 111, 123, 134

Tanakh 165–6
Taoist 247
tautology 50
teleology 87, 89, 90
temperature 24, 32, 35–6, 55, 116, 243, 274
tetragrammaton 165
Thackray, J. 251
theism 98–101, 111, 123, 128, 133, 155, 177, 183, 185, 187–9, 204, 212, 247, 254, 262, 266
theist 125, 135, 181, 196
theology 178, 184–5, 194, 196–7, 199, 211, 263
thermodynamic 32–3, 36–7, 43, 45, 273, 275–6
Thomas, R. S. 176, 197, 203–5, 261
tool 138, 191
transcendent 102, 122, 124, 137–8, 143, 162, 234–5
tree 49, 75, 89, 103, 113–17
Turing, A. 62
Turing machine 62

usefulness 7, 32, 239

value 7, 16, 18, 110, 121, 124, 131, 137, 141–2, 145–7, 151–2, 154–5, 177, 187, 227–8, 231, 246, 256, 259
van Inwagen, P. 185–6
variation 14, 46–8, 80, 86, 88, 98, 101–2, 105, 110
violence 68, 144, 159–61, 168, 253
von Neumann, J. 62

Ward, K. 100
Weil, S. 2, 256
whale 103
White, T. H. 169, 224, 268
Williams, R. 133, 143, 196, 203, 205, 212
willingness 69, 123, 152, 154, 156, 196, 211, 231, 254, 258, 267
wisdom 125, 129, 158–9, 165, 214, 247, 253, 267
witness 185, 201, 207, 211, 229, 250, 269
Wittgenstein, L. 196

zoology 18, 64, 82, 97